A social theory of innovation

ALEXANDER STYHRE

Liber

Copenhagen Business School Press

A social theory of innovation

ISBN 978-91-47-09773-9 (Sweden)
ISBN 978-87-630-0252-3 (Rest of the World)
© 2013 Alexander Styhre and Liber AB

Editor: Ola Håkansson
Cover and graphic design: Fredrik Elvander
Typeset: LundaText AB

1:1

Print: Sahara Printing, Egypt 2013

Distribution:
Sweden
Liber AB
S-113 98 Stockholm, Sweden
Tel +46 8 690 90 00
Fax +46 8 690 93 01
www.liber.se
Email: kundservice.liber@liber.se

Denmark
DBK, Mimersvej 4
DK-4600 Køge, Denmark
Tel +45 3269 7788
Fax +45 3269 7789

North America
International Specialized Book Services
920 NE 58th Street, Suite 300
Portland, OR 97213-3786, USA
Tel +1 800 944 6190
Fax +1 503 280 8832
Email: orders@isbs.com

Rest of the World
Marston Book Services, P.O. Box 269
Abingdon, Oxfordshire, OX14 4YN, UK
Tel +44 (0) 1235 465500
Fax +44 (0) 1235 465655
Email: client.orders@marston.co.uk

Author presentation

Alexander Styhre PhD is Chair of organization theory and management at the School of Business, Economics and Law at the University of Gothenburg. Styhre has published widely in the field of organization theory and management studies. His most recent books include *Visual Culture in Organizations* (Routledge, 2010), *Organizations and the Bioeconomy* (Routledge, 2012), and *Assembling Health Care Organizations* (co-authored with Kajsa Lindberg and Lars Walter, Palgrave Macmillan, 2012).

Content

Preface

Yannis Gabriel (cited in Suddaby, Hardy, and Huy, 2012) examines the management section of one of London's largest bookstores and notices that the entire range of management books is in partity with the Bauman-Bourdieu section in the sociology section, that is, the selection of management books is quite limited. Worse still, most of these management books are textbooks rather than research monographs. Suddaby, Hardy, and Huy (2012) ask whether this relatively limited commercial interest in management studies and management theories is indicative of the sparse development of new theories in management studies and organization theory? When Suddaby, Hardy, and Huy (2012) called for papers for a special issue of the *Academy of Management Review* on "New theories of organization", they only received commentaries on *how* such new theories should preferably be developed. Naturally disappointed with this outcome – not even when the most prestigious conceptual journal in the field had explicitly asked for new and innovative thinking were there any signs of such creativity – Suddaby, Hardy, and Huy (2012) suggest that there might be a need for longer, more integrated accounts in book format rather than short and condensed research articles in order to develop new and comprehensive theoretical perspectives. In the recent publication mania sweeping through academy, it is frequently pointed out that books never really count in the academic career race – only double-blind review journal papers have any merit and value. While this is not of necessity the whole truth, the research monograph is still an underdeveloped vehicle for theory development in management. It offers the space needed to fully develop an argument and it is not restrained by the strict formalistic structure of the academic journal article. However, research monographs can also represent many very different things; they can be constructions sites where new thinking is brought together or they can present final and integrated statements regarding insights already acquired; a kind of report about lessons learned. No matter how the book is used, in many cases – at least for me – it is a kind of platform (not entirely different from the platforms used in the automotive or pharmaceutical industries) upon which different research problems can both be articulated and integrated. The value of the book thus lies, perhaps, not so much in its final manifestation of printed hard-copy as in what happens during the writing process itself, and subsequently. Thus, all books are what Jacques Lacan called *poubellications* – a play of words on the French word *poubelle*, "garbage-can" and publications – kinds of waste-products in a non-pejorative sense of the term. For some, that is a terrible thing to say; for me, it is evidence of the value of the research monograph in intellectual pursuits.

This book is a kind of by-product of, or pendant to, studies I have been engaging in over the last few years in the field of life sciences. During this work, I have come across a number of scientists and researchers who have in detail accounted for the difficulties, uncertainties, and ambiguities that have riddled their work, testifying to the wide array of challenges attached to any scientific pursuit. Contrary to the popular image of the technosciences, these highly skilled and specialized professionals did not overlook the difficulties involved in their work, or underestimate the time-frames needed to bring, for instance, stem cell therapies to market. In addition, these interlocutors in many cases accounted for the importance of having a joint mission, a sense of camaraderie and shared objectives, including the creative skills and innovative thinking needed to accomplish anything of scientific value. Taken together, what these interlocutors expressed – or at least what I thought they did, potentially projecting my own preconceived ideas on them – was a concern regarding scientific work, a narrow and instrumental activity operating on the basis of extrinsic motivation and monetary rewards. In their narratives, scientific work and scientific career unfolded as a particular form of *modus vivendi*, a form of existence intimately bound up with theories, laboratory practices, and epistemic objects – widely recognized or in-the-making – whereby work could never be fully understood simply as something that puts food on the table, but as a lifelong intellectual engagement. Putting it another way, the life world of the scientist appeared to be transcending the narrow focus on the accumulation of economic resources, symbolic credentials, and reputation predicted by utilitarian theories of economic agents. The scientists' principal interests still seemed to extend beyond such narrow concerns; their perspectives reached into the future and they seemed to be endowed with capacities for foresight and ambition. As a consequence, I thought it might be of interest to reflect on the analytical tool to hand in the social sciences and organization theory more specifically in order to understand this way of living a scientific life. During a period when the field of expertise known as management is being brought into virtually every social sphere, it would be interesting to consider which images of agency and rationality are being enacted in the social sciences and organization theory. This book is, thus, an attempt to say something about agency and rationality in organizations in connection with innovation and creativity.

Topologically, the text is embedded in a Lacanian structure whereby play loosely corresponds to *the imaginary* (with play constituting and setting the boundary of the subject), reciprocity corresponds to *the symbolic* (with reciprocal relations constituting the self as a social actor), and squandering corresponds to *the Real* (waste and destruction being inherent in all production, yet commonly failing to be recognized as such, i.e., evading representation). As a consequence, there are more empirical studies that address playfulness and reciprocity while squandering, like Lacan's Real, essentially defies representation and is complicated to study empirically; the line of demarcation between waste and non-waste is a cultural and, ultimately, a political matter. Having said that, it is not, in my view, possible to attain a comprehensive and "final" social theory of innovation; instead, all such projects are bound to become fragmented and permeable analytical frameworks. Similar to the life sciences, examining the living but still failing to formulate a theory of what life is *per se*, theories of creativity and innovation examine the *outcomes* of intricate social processes but do not properly theorize how innovations are accomplished. Nevertheless, "weak frameworks" such as these may also be useful in pursuing thinking that helps to make sense of how economic value is created and perceived during the contemporary period.

* * * * *

I would like to thank a series of colleagues and collaborators, as well as acquaintances making contributions to my/our work: Kajsa Lindberg, Lars Walter, Rebecka Arman, Ola Bergström, Ulla Eriksson-Zetterquist, Andreas Diedrich, and Björn Remneland-Wikhamn at the School of Business, Economics, and Law at the University of Gothenburg. Ola Håkansson, a commissioning editor at Liber, has been a great help in supporting this book project. Finally, I would like to express my gratitude to the Bank of Sweden's Tercentenary Fund, for providing a research grant that enabled the empirical work preceding the ideas developed in this setting.

Sävedalen, May 2012

Alexander Styhre

<table>
<tr><td>Spiegel:</td><td>And what now takes the place of philosophy?</td></tr>
<tr><td>Heidegger:</td><td>Cybernetics.</td></tr>
</table>

Martin Heidegger: "Only God can save us":
The *Spiegel* Interview (1966)

Introduction

Economic and sociological perspectives on agency

> "'The axioms of a mathematical axiom-system
> ought to be self-evident.'
> How are they self-evident then?"
>
> *Ludwig Wittgenstein (1953: 223)*

The starting point of this book is that innovation is a social and collective accomplishment involving humans, artifacts, biological specimens, and other resources. Innovation work is situated practice bound up with both human ingenuity and the materiality engaged in the pursuit of innovation. Innovations are fabricated, made, and happen, but they are always bound up, one way or another, with human agency. Human agency, in turn, is both constituted and legitimated by certain rationalities and professional and social ideologies. Based on this elementary, perhaps even trivial, proposition, a social theory of innovation will be articulated.

The great mathematician-turned-philosopher and (now in hindsight) unlikely Harvard Business School professor Alfred North Whitehead once pointed to the need to combine "facts and imagination" in scientific work, a condition that unfortunately leads, in Whitehead's view, to the tragedy of making scientific work a most difficult pursuit: "Imagination is not to be divorced from the facts: it is a way of illuminating the facts … The tragedy of the world is that those who are imaginative have but slight experience, and those who are experienced have feeble imagination" (cited in Bennis and O'Toole, 2005: 102). Accounts of scientific work in its contemporary, highly disciplined, and institutionalized form at times overlook or ignore the imaginative component of research work, i.e. making scientific work less messy and non-linear than it actually is. Rheinberger (2010) praises Gaston Bachelard for freeing epistemology and science studies from the belief in symmetry, neatly ordered structures, and linearity, and for paving the way for perspectives that recognize the actual conditions under which science is produced: "Knowledge production continuously occurs in untidy places: in places of confusion and stubborn, ready-made opinions; in places where new things are tried out … unlike many of his contemporaries, Bachelard does not want to wall this untidy space off from epistemology: he proclaims it epistemology's heartland" (Rheinberger, 2010: 32). In addition, scientific know-how is not produced on the basis of rigorous procedures of verification and falsification; however, as Astley (1985: 504) remarks, "old paradigms fall from grace not because they are wrong but because they are boring". In Astley's (1985: 504) view, "successful scientists" are "intuitively perverse" and "always ready to question accepted world views and create opportunities for the critical rejection of what is taken as given by others". The works of Bachelard and Ludwik Fleck on epistemology, written during the interwar period, are precisely the accomplishments of such imaginative thinkers, capable of shifting the focus from "science-as-theory" to "science-as-experimental-practice", i.e., recognizing the *process* of scientific work rather than its output or end-point. In this view, science does not present itself as a series of axioms, theorems, theories, constants, coefficients, and so forth, but as a series of events and encounters between scientists, theoretical frameworks, and equipment and material resources put to use. Scientific work unfolds as a social practice, devoid of definitive starting points and with no stipulated termination points. The sciences are journeys to nowhere propelled by a combination of human curiosity and practical interest. What has been less emphasized in this shift of epistemology is the loss of the *Grund* or *Ursprung* (two German terms in Heidegger's thinking, here serving a purpose) of scientific work, the concern for and interested in which originally ignited the investigation into an area of interest, and the shift to what is done under the auspices of the sciences. Rather than being concerned with genealogy, the science-as-experimental-practice literature examines ongoing engagements and activities, and their practical implications. Ian Hacking (2010) has expressed his contempt for the notion of identifying or revealing "things primal", the essence and origin of things, the true inner nature of being or the first sparkle of an existence:

> I distrust the lust for things primal. There is, for sure, an ineluctable drive in Western consciousness to find the first moment. To find the skeleton of the first human being. To scan the first three seconds of the universe. To reveal the primal scene. Or to relieve the primal scream. Each of those programs makes sense, although some may prove to be wrong-headed or illusory. (Hacking, 2010: 67)

The works of Bachelard and Fleck rest on similar frameworks, not treating the sciences as a great chain of being from the *Ursprung* to the contemporary and differentiated modes of thinking, from "ancients to moderns", but as a social practice in a garden of forking paths, constantly dividing itself into new scientific procedures. In the new epistemological view of the technosciences, the assignment is no longer – using Clifford's (1997) metaphors – about finding the *roots* but the *routes*, the itineraries of the sciences. As a consequence, in such a research program, there is a skepticism against all claims regarding universalism, that human being or societies possess such and such qualities or regulatory mechanisms. One such universalist claim is the notion of human beings as autonomous agents – the *homo oeconomicus*[1] – capable of making individual decisions and having personal preferences and beliefs detached and isolated from the wider social setting wherein they live their lives. The calculating and self-interested human being is a useful heuristic in thought experiments, but mistaking it for a blueprint for social action would be a fallacy.

Homo oeconomicus and rational choice theory

The concept of *homo oeconomicus* has been, to say the least, controversial in the social sciences. John Stuart Mill defined *homo oeconomicus* as "a being who desires to possess wealth", while abstracting from "every human passion of motive" (Milonakis and Fine, 2009: 31). Being a conceptual model embedded in what economists call methodological individualism,[2] *homo oeconomicus* largely shields off the surrounding society in the enactment of agency. Thorstein Veblen spoke of *homo oeconomicus* in dismissive terms as the "hedonistic conception of man" (Milonakis and Fine, 2009: 161). Perhaps *homo oeconomicus* can be defined in negative terms as a rational and wholly self-regarding actor for whom "[c]ostly altruism is impossible *ex hypothesi*" (Healy, 2004: 389). In addition to rational choice theory, *homo oeconomicus* is perhaps the single most criticized term in the social science literature theorizing agency. Like no other concepts, perhaps, *homo oeconomicus* and rational choice theory induce the knee-jerk reaction that these are fallacious models of human thinking and acting, reducing

---

1 In this book, the spelling *homo oeconomicus* is used because it is closer to the Greek term *oikos* forming the root of a series of terms. Some of the texts cited use the spelling *homo economicus* and, in order to cite these correctly, the two standards will at times be mixed.

2 Economists distinguish here between "institutional individualism" and "psychological individualism" where the former perspective assumes that the individual subject is guided by institutionalized norms and beliefs while the latter perspective, used in neoclassical economic theory and marginalist theory, emphasizes individual preferences. Institutional individualism is thus closer to the social science enactment of agency inasmuch as it recognizes shared social beliefs as guiding and directing human action.

human existence to a series of selection choices and calculative practices.[3] It is
thus important to recognize that, as theoretical models, these kinds of thought
experiments have great value when it comes to theorizing how economic decisions are made under determinate conditions. Boudon (2003) emphasizes, with
reference to Howard Becker, that a theory holds less explanatory value as soon
as it invokes "external forces" such as psychology, biology, and culture as factors
explaining a particular behavior or phenomenon because "all these explanations
raise further questions" (Boudon, 2003: 3). Explanations cannot make use of too
many *ad hoc* hypotheses or seek legitimacy outside of the propositional framework constituting the theory. Theories need to be *economical*, i.e., to be condensed, and make use of what is to hand. In that respect, rational choice theory
is an elegant theory inasmuch as it locates the explanatory value in individual beliefs and preferences.

Hechter and Kanazawa (1997) list a number of research fields in sociology, including studies of family and demography, religion, gender, crime and deviance,
and stratification and mobility, which have all made use of rational choice theory.
Thus, various social conditions are at least partially explained on the basis of rationalist theories; there is no iron curtain between social and economic matters
in everyday life, but they tend to be bound up with one another (see also Linden-

---

3 The term *homo oeconomicus*, "economic man," is etymologically derived from the Greek
terms *oikonomia* and *oikonomikos*, meaning the "householding of resources". As emphasized by
Mondzain (2005: 22–34) and Agamben (2009: 11), this term has strong theological connotations
and, consequently, the term *homo oeconomicus* is riddled with inconsistencies between the historical and contemporary uses of the term "economy". Mondzain (2005: 18) says that the term
oikonomia does not appear in the works of Homer, Hesiod, Herodotus, or Thucydides, instead
first being used by Xenophon and, eventually, Aristotle. In these authorships, "economic discourse is a logos that lends its epistemological status and purpose to a mediation on the administration and management of domestic life, specifically the philosophical and practical consideration of the management of private fortune, particularly in the rural domain" (Mondzain, 2005:
18). Very quickly, this concept was used to denote similar procedures of administrating goods
in the public domain. In the medieval and scholastic theological discourses, the terms *oikonomia* and *oikonomikos* were eventually appropriated in order to denote the orderliness of God's
creation: "As in Aristotelian thought, the church fathers believed that the principles of organization, management, and administration should be drawn from a natural model, which is in this
instance a divine model. This is because God is the *oikonomos*, the supreme administrator and
manager, and the ensemble of his creation in the universe is *oikonomia*" (Mondzain, 2005: 35).
This theological framework was eventually drawn on when the "natural philosophers", e.g. Caro-
lus Linnaeus (more widely known in Sweden as Carl von Linné), articulated their theories about
the constitution and organization of natural and biological systems. Says Gammon (2010: 223):
"When political economy arose in the eighteenth century, it took its notion of 'economy' not from
the market as a pre-existing social institution, but from the belief in a broader divine natural
economy. 'Oeconomy' was a word with wide circulation, commonly referring to matters of physiology." The term "economy" has thus been used historically not so much to denote individual calculative reason but as a broader theological model of the cosmos as a thoughtful and inherently
reasonable arrangement under the directorship of God. In speaking of *homo oeconomicus*, economists thus separate the term from its etymological roots. Rather than emphasizing individual
agencies and interests, the term *oikonomia* captures being in its full existence.

berg and Frey, 1993). Boudon (2003) outlines the postulates that constitute rational choice theory:

> RTC [Rational Choice Theory] can be described by a set of postulates ... The first postulate, P1, states that any social phenomenon is the effect of individual decisions, actions, attitudes, etc. (individualism). The second postulate, P2, states that, in principle at least, an action can be understood (understanding). As some actions can be understood without being rational, a third postulate, P3, states that any action is caused by reasons in the mind of the individual (rationality). A fourth postulate, P4, assumes that these reasons derive from consideration by the actor of the consequences of his or her actions as he or she sees them (consequentialism, instrumentalism). A fifth postulate, P5, states that actors are concerned mainly with the consequences to themselves of their own action (egoism). A sixth postulate, P6, maintains that actors are able to distinguish the costs and benefits of alternative lines of action and that they choose the line of action with the most favourable balance (maximization, optimization). (Boudon, 2003: 3–4)

It is easy to see what makes rational choice theory appealing to theorists from various disciplines and camps: it is a neat and elegantly structured theory that is capable of explaining a long series of empirical phenomena. Still, as Boudon (2003: 6) makes clear, although it is a powerful theory, it nonetheless appears "[p]owerless when confronted with many phenomena" (see also Collins, 1993). A substantial list can be compiled of familiar phenomena that the theory cannot fully explain (e.g., voting in democratic societies, gambling, and religious beliefs). For sociologists and other social theorists inclined to make reference to social structures and communities as explanatory factors, rational choice theory is guilty of confusing, primarily, postulates three and four in Boudon's model. That is, rationality is taken to be exclusively concerned with instrumental reason. If there is an underlying rationality in an actor's behaviors, embedded in beliefs, norms, ideologies, and identities, then that rationality will not always be concerned already with accomplishing instrumental goals. In Boudon's (2003: 17) phrasing, "[r]easons dealing with costs and benefits should be given more attention than they deserve. Rationality is one thing, expected utility another." Boudon (2003: 17) concludes: "On the whole, to get a satisfactory theory of rationality, one has to accept the idea that rationality is not exclusively instrumental. In other words, the reasons motivating an actor do not necessarily belong to the instrumental type."

In the field of neurology, too, studying the human brain on the basis of scientific, quantitative methods, representing an entirely different epistemology than mainstream sociology, the instrumental and linear view of human cognition and decision-making has been criticized. "[I]t has become abundantly evident", argued two neuroscientists, in the introduction to a special issue of the journal *Neuron* published in 2002 (cited by Schüll and Zaloom, 2011: 518), "that the pristine assumptions of the 'standard economic model' – that individuals operate as optimal decision makers in maximizing utility – are in direct violation of even the most basic facts about human behavior." Schüll and Zaloom (2011: 518) continue:

> Grounding their research in empirical data, either constructed in a laboratory or drawn from more naturalistic settings, these subfields have documented countless instances where human behaviour does not seem to follow the laws of rational economic action; on the contrary, their data has shown that people systematically depart from such laws when they weigh information and judge probabilities.

When the propositions regarding the human behavior guiding economic theory and policies derived from that lend themselves to empirical investigations in the "hard sciences", there is a mismatch between theoretical frameworks and observed behavior and empirical findings. "What economists mean by rationality is not exactly what most people mean. What economists mean is better described as *consistency*", says Joseph E. Stiglitz (2010b: 249), a Nobel Prize laureate in Economic Sciences. He continues: "Research over the last quarter century has shown that individuals do act consistently – but in ways that are markedly different from those predicted by the standard model of rationality. They are, in this sense, predictably irrational" (Stiglitz, 2010b: 250). As another Nobel Prize laureate in Economic Sciences, Daniel Kahneman (2011), has clearly demonstrated, even in cases where individuals are given two distinct choices – a typical experimental situation in psychological research emphasizing individual decision-making and calculable alternatives – these individuals have great difficulty avoiding fallacies in order to make "rational decisions". The underestimation of risk in certain situations, while overrating them in other cases, the inability to draw conclusions strictly based on available data, and prejudice, are just a few psychological factors misleading individual decision-making. At the same time, as Beckerts (2002: 13–14) remarks, bearing Weber's concept of ideal-types in mind, there is a use for a "purely heuristic construct" to be able to construct theories, but "[s]uch a position is always confronted with the obligation to justify why *precisely this* type of action is shifted into a privileged position, as opposed to which all other types of action orientation can constitute only a residual category". That is, it is not the idea of *homo oeconomicus* per se as an ideal-type that is the problem, but how this model has acquired a hegemonic position when modeling and explaining economic action. For Beckerts (2002: 17–18), then, the role of economic sociology does not so much aim to debunk the construct of *homo oeconomicus*, but to theorize the *social embeddedness* of economic action; that is, to theorize economic agency without assuming what Beckerts calls "a transcendence of selfish objectives".

Efficient markets and other hypotheses

It may be helpful to postulate a general theorem that human action is guided by rationality, but restricting such rationality to strictly calculative reason, a narrow-minded "costs and benefits analysis" would be to reduce the life world of human beings to a strictly economical level (see, for instance, Tyler [2011] on the research into cooperation). The scientific discipline of economics is often criticized by economic sociologists – and by economists for that matter; see, for instance, McCloskey (1986) – for advocating theoretical perspectives regardless of their practical accuracy and for conspicuously ignoring research evidence produced within the social sciences. "Economics is supposed to be a predictive science, yet many of the key predictions of neoclassical economics can easily be rejected", says Stiglitz (2010b: 245). Even economists themselves, at least outside of mainstream and formalist circles, may deplore the weak connections with the other social science disciplines and the humanities, the most noteworthy being history. "[M]ainstream economics, trapped between perfecting their increasingly esoteric and formalistic models and techniques, no longer show any interest in anything that lies outside their mode of thinking and their field of competence, including the history of their own subject and the methods employed", write Milonakis and Fine (2009: 9). Undoubtedly, economics has advanced its prestige and position in contemporary society and politics is increasingly being subsumed under the regime of economic reason and economic theory (Fourcade, 2006, 2009; Kogut and Macpherson, 2011). Today, the political elites are increasingly being recruited from the professional community of economists. As a consequence, there are many reasons for social scientists to be critical of the expanding jurisdictional domain of economists. Management scholar Jeffrey Pfeffer (1993) has officially declared that he wants the field of management to play a similar role in society to economics. Economic sociologists are critical of the underlying theoretical assumptions made during the construction of theories. Perhaps the most important critique suggests that, rather than being a scientific endeavor capable of *describing* economic conditions and predicting possible outcomes, economics serves a *performative function* in providing the script for economic action (Mackenzie, 2004, 2006). For instance, the so-called *efficient market hypothesis*, one of the central propositions in economics, is more of an assumption than an empirically-proven fact, however.[4] Still, the theory played a key role in regulating investment behavior in the 1990s; Mizruchi and Brewster (2005: 301–302) suggest that: "During the 1990s bull market, efficient market theory played an important role in boosting stockholder confidence. As stock prices rose dramatically, many investors chose to stay in the market, based on the argument that stock prices reflected firms' true value." Economics is not, then, a science in the conventional sense of the term, i.e. exploring and explaining mechanisms and causal relations existing independently of human beliefs and action, rather it sets up a number of rules to be followed by human agents and therefore, *ipso facto*, makes the prescribed theories come true via its capacity to coordinate and align actions. Economic theory is not the mirror of economic reality but its blueprint.

---

4 The so-called *efficient market hypothesis* is associated with the work of Eugene Fama (e.g., Fama, 1970) and colleagues at the University of Chicago and is commonly associated with three forms. Mizruchi and Brewster (2005: 292) explain: "The weak form suggests that prices efficiently reflect all the information contained in the past series of stock prices. It is therefore impossible to earn superior returns simply by looking for patterns in stock prices. The semistrong form suggests that prices reflect all published information. This means it is impossible to make consistently superior returns simply by reading the newspaper, looking at the company's annual accounts, and other public information. For this reason, analysts can do little to help an investor earn superior returns. The strong form of the efficient market hypothesis (the one the most popular in the 1990s) suggests that stock prices effectively incorporate all available information: the consequence of million of investors competing for the edge is that virtually no source of information remains unexplored." Mizruchi and Brewster (2005: 292) review a number of empirical studies in both economics and economic sociology demonstrating that the efficiency of the market hypothesis is poorly supported by data. Hayward and Boeker (1998) found that, contrary to the efficient market hypothesis, securities analysts rate their own bank's clients' securities higher than other companies, indicating that not all information is available to all actors. Zuckerman (1999) demonstrates that stocks subject to shared classification are, everything else being equal, more highly valued than stocks that are not clearly classified. Market analysis thus becomes an interpretative endeavor demanding extra resources. "For a product to compete in any market, it must be viewed by the relevant buying public as a player in the product categories in which it seeks to compete … [s]uccess or failure at gaining such recognition has a significant impact on a firm's fate in financial markets. All other things held equal, firms that cultivate an egocentric network of reviews to securities analyst that reflects its industrial participation are more highly valued than those that do not", contends Zuckerman (1999: 1429). As a consequence of such ambiguities in financial markets, there are many intermediary actors, such as rating agencies (Fleischer, 2009), that play a key role in handling market inconsistencies: "Armies of interpreters and prognosticators are present on Wall Street because they fill an important social purpose: they help investors make sense of the dizzying array of possible investments. No such investment has a clear value, and the struggle to anticipate future prices never ends" (Zuckerman, 1999: 1431). In addition, Zuckerman (2004), Zajac and Westphal (2004), and Uzzi (1999) provide evidence that markets are determined by social factors, turning the efficient market hypothesis into what Kogut and Macpherson (2011: 1309) call an *ideological innovation*, a proposition that is "[e]mbedded within a larger program to achieve a desired state of the world".

On the performativity of economic theory and the creation of new ideas

British sociologist Andrew Pickering (1995) makes a distinction between a *representational* and a *performative* idiom in scientific work. The representational idiom, argues Pickering (1995: 5), "[c]asts science as, above all, an activity that seeks to represent nature, to produce knowledge that maps, mirrors, or corresponds to how the world really is." In pursuing such representations, continues Pickering, "it precipitates a characteristic set of fears about the adequacy of scientific representations that constitute the familiar philosophical problematics of realism and objectivity." In contrast, the performative idiom of science is "[r]egarded a field of powers, capacities, and performances, situated in machinic captures of material agency" (Pickering, 1995: 7). The performative idiom does not aim to mirror external realities, rather it helps to produce certain social effects. Elsewhere, Pickering (2010) talks about a *performative epistemology* as something capable of producing reality. Callon (2007: 315) makes a similar distinction between *constantive* and *performative* statements in science: "Scientific theories, models, and statements are not constantive; they are performative, that is, actively engaged in the constitution of the reality that they describe." "The concept of self-fulfilling prophecy seems to apply to economics. Economics – and this is where it derives its strength – is a constructed, logical discourse based on a number of irrefutable hypotheses", contends Callon (2007: 322). In the economic sciences, there are many examples of the performative idiom, the joint mapping and constitution of economic agency. In accounting, Hines (1988: 258) has argued that the "financial accounts of organizations" do not merely "describe or communicate information about, an organization, but how [they] also play a part in the construction of the organization, by defining the boundaries" (see also Robson, 1992). Accounting is not, then, merely a disinterested mapping of financial assets and their use but also actively constructs the very notion of the organization as a portfolio of financial assets and resources. Another case of performativity is the so-called Black-Scholes-Merton option pricing model examined by MacKenzie and Millo (2003). MacKenzie and Millo (2003) argue that, originally, the model was "theoretical rather than empirical" and when it was tested in 1972, it could not accurately predict option prices. As financial traders were in need of a model to guide their decision-making as regards whether or not to buy financial assets, they started to use the pricing model regardless of its initial shortcomings. The first step included a modification of the financial markets and the procedures for trading:

> [The] empirical success was not due to the model describing a preexisting reality; as noted, the initial fit between reality and model was fairly poor. Instead, two interrelated processes took place. First, the market

> gradually altered so that many of the model's assumptions, widely un-
> realistic when published in 1973, became more accurate. (MacKenzie
> and Millo, 2003: 122)

Second, the model was increasingly used to guide trading and, as an increasing
number of traders started using the same calculative practices, prices tended to
converge toward the "theoretical values" of the options. "Gradually", say Mac-
Kenzie and Millo (2003: 127), "'reality' (in this case, empirical prices) was performatively reshaped in conformance with the theory." The Black-Scholes-Merton
option pricing model, at times put forth as the jewel-in-the-crown of neoclassical
economic theory, allegedly testifying to the practical usefulness of economics,
was initially an inaccurate model for predicting option prices; however, as the
community of financial traders started to use it, it "became true". This is indicative of the performativity of economic theory – its ability to produce self-fulfilling
prophecies. Such social effects are not problematic *per se* in the eyes of external
analysts but they poorly blend with the "science as representation" idiom – when
obeying what has been called the *law of non-contradiction*, either science per-
forms social effects *or* it mirrors underlying relations and conditions, not both.
When economics claims it is in a position to be able to make predictions about
economic outcomes, without recognizing the performative powers of the articulations of the predictions, it transgresses the boundaries between the two scientific idioms. Mackenzie (2006) points to the value of economic and financial theories as a set of heuristics cognitively supporting actors in their day-to-day work:

> Part of the significance of the performativity of economics arises from
> the fact that individual human beings tend to have quite limited powers
> of memory, information-processes, and calculation. Those limitations
> explain the centrality – especially in complex markets – of simplifying
> concepts (e.g., 'implied volatility') and of material means of calculation
> … The concepts and material means are therefore constitutive of eco-
> nomic action … [which] implies that the economic theory crystallized
> in concepts and devices is performed. Sometimes, I have suggested,
> this performativity is Barnesian [after the sociologist of science Barry
> Barnes]: it helps to create the phenomenon the theory posits. (Mac-
> kenzie, 2006: 265)

Similarly, McCloskey (1986: 175) argues that economics legitimates itself through
the use of persuasive rhetoric: "The main achievement of economics is not the
prediction and control assigned to it by modernist social engineering, but the
making sense out of economic experience." Economics is, then, accused of being "impure" (in Douglas' [1966] sense) in terms of blending and mixing the two
scientific idioms. The sciences either represent reality (as in the chemistry of mo-

lecular biology) or they are performative (as in the case of management sciences and accounting), but they must not claim to do one thing and *de facto* do the other. Sciences are either *maps* or *tools*, not both. This is the first critique of economic theory and economics.

The second critique is the elimination or disregard of what may broadly be labeled as "the social"; for the sociologist, a sacred domain capable of explaining most societal phenomena. Pierre Bourdieu has been particularly animated in his critique of such a reductionist view of economic agency. For economists, sociology is a "pariah science", claims Bourdieu (2005: 10). In his view, economics is "disembedded", separated from "the social order": "The science called 'economics' is based on an initial act of abstraction that consists of dissociating a particular category of practices, or a particular dimension of all practice, from the social order in which all human practice is immersed" (Bourdieu, 2005: 1). That is to say, rather than assuming that "the most basic economic dispositions" (Bourdieu, 2005: 8), e.g. needs, preferences, and propensities, are exogeneous, i.e., "dependent on a universal human nature", these are, says Bourdieu, *"endogenous* and dependent on a history that is the very history of the economic cosmos in which these dispositions are required and rewarded." That is to say, rather than making the assumption that economic agents are equipped with a certain set of preferences and beliefs, such preferences and beliefs are in themselves the effects or consequences of economic action; the deed produces the doer rather than the doer being an *a priori* enactment of human agency. Freese (2008) makes a similar remark, pointing to the differences between economics and sociology:

> Simple, uniform actors have had their greatest run anyway in orthodox economics, with its assumptions of uniform, unchanging preferences and actions as optimal given preferences … Sociologists traditionally look at the world and see actors who have biases, are swayed by emotions, respond to how information is framed, are influenced by affiliations and identity, internalize rules, and so on. (Freese, 2008: S21)

Bourdieu's rant regarding the assumptions made in economic theory can be supplemented by more modest critiques, e.g. Mackenzie's (2009) study of how financial markets are constituted by a combination of a variety of heterogeneous resources that includes finance theory, digital media such as computers and Internet connections, calculative practices, and communication technologies. In such a view of the financial market, it is complicated to separate "the technical" from "the social" as the two are mutually constitutive:

> The 'technical' and the 'social' are not two separate spheres, but two sides of the same coin, as a long tradition in the social studies of sci-

ences and technology has emphasized. Any market ... is a sociotechnical construction. A central role of the social studies of finance and similar 'material sociology' approaches to markets is thus to add to the well-honed set of tools for analyzing the more directly social aspects of markets ... a set of tools for making sense of their more 'technical' aspects. Because the 'social' and the 'technical' are inextricably linked in market construction, the two sets of tools will ultimately need to be integrated fully: a challenging but important academic task. (Mackenzie, 2009: 181)

No economic action subsists in a social vacuum; economic agency is fundamentally social and needs to be understood as such. The thesis being put forth here can be expressed accordingly: economic theory, and consequently management theory (in contrast to organization theory, being more indebted to and bound up with social science disciplines such as sociology and political science) is enacting a delimiting view of economic agency; economic agency is bound to instrumental rationalities and utilitarian thinking and overlooks a series of social conditions, institutions, and beliefs regulating social life and economic pursuits. This does not make the concept of *homo oeconomicus* and its accompanying procedures and rationalities an irrelevant or useless analytical model; apparently, there is a series of economic transactions and decisions being made on the basis of strict instrumental calculations. If an individual can choose between receiving an interest rate of 3 per cent on one bank account and 4 per cent on another, *ceteris paribus*, then he or she will be likely to choose the second alternative. However, many economic activities are surrounded by uncertainty, ambiguities, competing interpretations of the event, and so forth, making many economic decisions and activities more complicated to deal with, i.e., the clause *ceteris pari-*

bus is not an unproblematic assumption.[5] This is precisely why sociologists like Pierre Bourdieu (2005: 197) are critical of economists like Gary Becker (1992) – "the leading proponent of … economics imperialism" (Milonakis and Fine, 2009: 306). The questions of whether or not, for instance, to marry or to have children are, to some extent, economic ones (the number of children being born falls during periods of economic turmoil, for instance) but cannot be *reduced* to strictly calculative decisions *ex ante* since such decisions are made within a dense social texture prescribing certain standardized life choices (e.g., get an education, find a spouse, become parents, eat your vegetables, live a long and happy life, and die peacefully at a statistically predictable age).[6]

---

5 The very notion of *homo oeconomicus* is based on the scientific ideal of *precision*, the capacity to produce exact measurement expressed in numerical form. This scientific ideal has its roots in Victorian Britain, suggests Schaffer (1995), a society characterized both by substantial social change (e.g., industrialization, urbanization, growing class conflict) and the emergence of the social sciences (sociology, criminology, political science) as a form of social engineering. "The statistical investigation of poverty, crime, education, sanitation, and employment of working people that became almost an obsession in Britain, beginning around 1830 reflected the greater distances between classes in an industrialized society", argues Porter (1995: 192). In a society where class privileges were questioned and no longer served as a legitimate source of authority, professional work increasingly had to rely on technical and scientific expertise. Consequently, precision became a highly-valued quality for, for instance, the actuaries working in the insurance industry (Porter, 1995). The ability to produce numbers and ratios was regarded as a legitimate way of knowing the world and making factual statements about it: "In a large-scale industrialized society, numbers provide an appropriately impersonal way of learning about people in alien settings", says Porter (1995: 192). The use of calculation machines and other tools for exact calculation did not, however, initially evolve in the sciences but in commercial and bureaucratic and administrative activities (Warwick, 1995: 313) and only in due course were the sciences expected to present exact measurements like those used in financial calculations and budgets. Again, the advancement of the sciences as calculative, mathematical procedures is due to the world of commerce (Rotman, 1987). The concept of *homo oeconomicus* is thus grounded in scientific ideologies, ideologies which in turn have imported calculative practices and epistemological ideals concerning precision from commercial and bureaucratic endeavors, pointing at an interesting genealogy of the concept, a form of the co-production of scientific ideologies and practical calculative procedures. The "economic", i.e., calculative man is historically contingent, emerging within a specific historical setting wherein the relations between the social classes are negotiated.

6 Daniel Kahneman (2011: 412) offers a fine illustration of the implications of Becker's enactment of human agency: "I once heard Gary Becker … argue in the lighter vein, but not entirely as a joke, that we should consider the possibility of explaining the so-called obesity epidemic by people's belief that a cure for diabetes will soon become available." One might think that many scholarly professions in medicine, sociology, and social work would not be too ready to explain the growth of obesity as a matter of a conscious choice based on calculated risks without taking into account wider social changes and conditions (see, for instance, Monaghan, Hollands, and Pritchard, 2010; Vrecko, 2009; Throsby, 2009). Becker's theoretical claim, i.e. that investment in human capital is a rational and strategic choice that will pay off as markets price assets and skills correctly has not been substantiated by empirical evidence. On the contrary, research published by, for instance, Castilla (2008) into what he calls the *performance-reward systems* of an internal labor market at a major American corporation show that such systems disfavor women and minorities. What Castilla (2008) refers to as the *performance-reward bias* thus undermines Becker's theory of human capital investment: Certain groups do not receive rewards on a par with their human capital investments.

Especially when it comes to the issue of creativity and innovation, i.e., social and organizational renewal in the main, the instrumental rationalities and utilitarian thinking conspicuously fall short. Innovation is embedded in social relations and joint collaboration. "Innovation is just as social as conformity", says Charles Horton Cooley ([1902]1964: 41), making reference to Gabriel Tarde (1890] 1962), the great French sociologist who examined the processes of innovation and imitation (Barry and Thrift, 2007). "Invention, even more than science, is a social phenomenon; in quite matter-of-fact ways, it is a human activity which can only be fulfilled when certain social conditions obtain", say Burn and Stalker (1961: 21) in their classic study of innovation in Scottish firms. Innovation is a social and collective process. "The notion of the hermit genius, spinning inventions out of his intellectual and psychic innards, is a nineteenth-century myth", declare Burn and Stalker (1961: 21).

The creation of new ideas, or the combination of old ones, the primus motor of the capitalist economy in the Schumpeterian tradition of thinking, cannot be explained on the basis of reductionist analytical models centered on individual agents. As a consequence, this book will discuss three principles supporting and complementing instrumental rationalities. These are:

▶ The principle of *playfulness*, recognizing the need to operate outside of strictly rule-governed regimes of control.

▶ The principle of *reciprocity*, suggesting that all human beings are located in social settings where they give things away for free in order to receive new artifacts (gifts), ideas, information, and recognition.

▶ The principle of *squandering*, underlining the need to systematically waste accumulated resources in order to jointly constitute and reproduce specific social formations.

In various ways, these three principles enact human and economic agency as something that includes both the instrumental and calculative rationality of the *homo oeconomicus* model and other "extra-instrumental rationalities" – rationalities that extend beyond the short-sighted temporality of utilitarian thinking. In advancing such a view of economic agency, alternative rationalities are recognized that are potentially capable of shedding further light on the use of innovation and creativity by organizations. In strict ideologies dominated by instrumental rationalities, such principles may appear absurd, or downright irrational, but when one thinks carefully about how new ideas evolve and gain a foothold in recalcitrant societies, it becomes evident that new technologies and ideas are not strictly the outcome of instrumental thinking but are based on a variety of human faculties and capacities. Studies of, for instance, the telegraph (Israel, 1992; Israel, 1998), the telephone (Fischer, 1992), the bicycle (Bijker, 1995), the automobile (Franz, 2005), new communications and information media (Wajcman and Rose, 2011), robots and social media (Turkle, 2011), the synthesizer (Pinch and Trocco, 2002), and the gramophone (Faulkner and Runde, 2009) show that users play a key role in anchoring technological innovation in social settings, causing them to play a part in everyday life (Oudshoorn and Pinch, 2003). The argument is, then, that instrumental rationality is only capable of embodying a subset of these human faculties and capacities and that, consequently, human agency needs to be theorized outside of the relatively narrow sphere of instrumental rationality. When the sun rises over the totality of human accomplishments, it shines on all of humanity, not only on analytical and reductionist skills. Human societies are the outcome of a variety of events and conditions, but they are not solely understood on the basis of utility and personal gain but also on the basis of collective rituals and joyful expenditure. Economic agency, then, is to be examined along alternative routes.

A social theory of innovation

The title of this book, *A social theory of innovation*, deserves some clarification. As Bruno Latour has emphasized time and again, "the social" and "society" are not primordial structures capable of explaining human action. On the contrary, such terms themselves need to be explained and are constituted by human actions; "In most situations, we use 'social' to mean that which has already been assembled and acts as a whole, without being too picky on the precise nature of what has been gathered, bundled, and packaged together" (Latour 2005: 43). Society is constructed from the "bottom-up"; consequently, it is a fallacy to refer to "the social" or "society" in order to justify any qualitative claims regarding human action. On the other hand, terms such as "the social" have a polemical function inasmuch as they seek to capture the totality of human existence, and not just the subset of economic and calculative practices. In the title of the book, "the social" is put forth as a contrast to "economic", i.e., that which extends outside of the narrower domain of the particular form of rational thinking embedded in calculative reason. Second, the term "theory" is a contested concept. Håkansson (2007: 65) emphasizes that theory "[i]ncludes everything from inherited mental maps to scientific theorems", i.e., it is a relatively broad term ranging from detailed and closely knit ensembles of concepts anchored in empirical evidence to more sketchy perspectives and worldviews grounded in preferences, activist agendas, and ideologies (as in, for instance, "critical theory" or "feminist theory"). In a performative epistemology, theory does not so much mirror underlying factual conditions as form part of what Zahra and Newey (2009: 1059) speak of as "intellectual arbitrage" whereby theory "[g]ives meaning to data, defines phenomena, explains and interpret findings, and fuels discoveries". This is arguably a "weak" form of theory, not capturing essential conditions but rather acting as a heuristic or a tool mobilized and constantly modified during intellectual pursuits. Charles Wright Mills (1959) argues that there is difficulty attached to articulating what he referred to as a "grand theory" because it is complicated to bring down to earth and make operable in empirical studies: "When we descend from the level of grand theory to historical realities, we immediately realize the irrelevance of its monolithic Concepts", says Mills (1959: 44). On the other hand, Mills is sceptical regarding the claim that all theories need to be carefully supported by exact data and evidence: "We can of course, by suitable abstraction, be exact about anything. Nothing is inherently immune to measurement" (Mills, 1959: 73). In Mills' view, there is a relatively narrow field wherein theories may be interpreted, neither as part of a grand theoretical framework nor as strictly articulated on the basis of the data amassed. Mills does not use terms such as performative epistemology but proposes that theories are primarily "tools for thinking", joint conceptual frameworks that help professionals discuss certain issues and perceived problems. The key epistemological issue here is that there is always a difference, a certain distance, a caesura, between what is actual and what is represented: "If representation were a transparent image of the represented, it would not even be thinkable, it would paradoxically disappear. Representation can exist only if there is a gap between the represented and the representer, if the transparency which representation aims is not achieved. A true representation is impossible and from this impossibility the possibility of representation emerges", argues Quattrone (2006: 150). Thus, theory, "[t]he cognitive frames of reference that enable a community to make sense of the messages exchanged within it" (Håkansson, 2010: 1810), is never a comprehensive or exhaustive mapping of the actual but remains a sketch of collectively enacted life worlds enabling further communication and exchanges. What the representer (researcher and theorist) represents (social and natural phenomena) can never be conclusive, but there needs to be a gap between the actual, i.e., what is represented, and the representation.

Swedberg (2012) suggests that the concept of *theorizing* may offer some reconciliation between what is represented and the representer; theorizing is the practice of formulating new theories on the basis of empirical data – Swedberg prefers the term "observations" – and thus occurs in what Swedberg speaks of, following Reichenbach (1938), as the "context of discovery". Swedberg points to the difficulties involved in creating new theory:

> The dilemma for much of contemporary social science is that you are damned if you do and damned if you don't. It is hard to produce good theory if you start from the facts; and it is hard to produce good theory if you start from theory. In the former case, there will be no theory; and in the latter case, the theory already exists. (Swedberg, 2012: 7)

In an attempt at resolving this dilemma, Swedberg (2012), supported by Charles Wright Mills, Durkheim and Weber, stresses the empirical data as the starting point. In Swedberg's view, theorizing is the creative and venturesome part of scholarly work, the processes whereby the researcher inevitably encounters his or her own potentials and limitations: "To theorize you need to open up yourself, to observe yourself, and to listen carefully to yourself. One's imagination, intuition, and capacity for abduction all need to be observed, reflected upon, and developed by each individual, in his or her own unique and personal way", says Swedberg (2012: 34).

Consonant with Swedberg's (2012) arguments in favor of theorizing, Flusser (2002: 86) suggests that a social theory of innovation is not so much an attempt at advancing a set of "fixed, unchanging ideas" as an "active modelling" of how, for instance, human beings are collectively – at times even individually – capable of producing new innovations. Similarly, Barad (2011: 451) argues that "theories are not mere metaphysical pronouncements on the world from some presumed position of exteriority". Instead, Barad continues, "theories are living and breathing reconfiguring of the world". Perhaps this theoretical pursuit shares a few characteristics with what Nayak (2008) calls *processual theories*:

> Theory is not about its object, but thinking beyond the object. The outcome of processual theory construction is not explanation. Nor is theory a way of writing that recognizes its constructed nature. Theory, in this sense, is not a process of moving towards a better representation of underlying realities … instead, it is pure creativity, always moving beyond itself. It *is* not but it *acts*; it is not an atemporal or abstract representation of reality but the active or the useful or, more precisely, the creative. The notion of mobility, movement and change are the primordial qualities of processual theory. (Nayak, 2008: 186)

If we understand Nayak (2008) correctly, theory is not so much a mapping of reality as the very process of making use of theory when seeking to understand human action and social conditions.

A social theory of innovation is not, then, a final and fixed framework, preferably modeled into what Lynch (1991: 11) calls "theory pictures", visual representations of abstract theories composed of "pictorial elements" such as "discretely positioned names, bounded figures, and connecting lines and arrows" serving as the "grammatical operators" of the theory picture. Expressed differently, a social theory of innovation is by no means a full-fledged, integrated, and empirically grounded theory of innovation as a social phenomenon ready to lend itself to empirical testing. Instead, it is a framework for understanding innovation grounded in both empirical studies and theoretical reflections. Like all theory, it is a patchwork that is constructed from two different starting points, as a theoretical engagement and empirical investigations. It is an attempt at "actively modeling" how innovations are produced, not so much as the effect of calculative rationalities but as a joint and inherently playful activity aimed at producing new ideas and materializations of such ideas.

Outline of the book

This book consists of this introduction and five chapters that follow it. In Chapter One, the discussion regarding economic agency is further developed and related to the field of organization theory and management studies, emphasizing the need to think about how creativity and innovation are produced in organizational settings increasingly concerned with reducing costs and shortening timelines. Chapter Two introduces the concept of play and playfulness as an important human capacity which is represented in all sorts of societies and which, arguably, is the principal driver behind new and innovative thinking. In Chapter Three, the concept of reciprocity, just like the concept of play, is observed in a variety of historical and contemporary societies, demonstrating a variety of mechanisms for reciprocal relations. Only by giving can humans receive, thus many economic endeavors are embedded in reciprocal relations. Chapter Four examines the procedures and mechanisms for squandering and wasting resources in various domains of society and organizations, pointing to the ritual and ceremonial value of waste as the recognition of the resources of collectively being accumulated. In the final chapter, Chapter Five, some of the arguments made are discussed, especially in relation to the perception of time, and the implications for studying organizations and management are addressed. The key argument here is that situational rationality operates along a variety of temporalities, some linear, some circular, and some on the basis of other geometries, and that instrumental rationalities are embedded in linear mechanical clock-time, a temporality that is at odds with the duration of the lived time lying at the heart of any creative and innovative thinking.

Needless to say, this is a syncretic book based on a transdisciplinary corpus of literature that includes organization theory, sociology, anthropology, philosophy, and on whatever has been found useful in pursuing the argument. At the same time, the book addresses a specific organization theory and management concern, that is how renewal and creativity can occur within a mode of thinking that carefully seeks to eliminate any slack or non-value adding activity, i.e., the puritanical monitoring of the uses of organizational resources in the economic regime that has been called "the quarter economy". For instance, how can biotechnology firms and pharmaceutical companies produce new therapies and explanatory models for the biological pathways being explored within the short-sighted temporal horizon of the venture capitalists? New contributions, new ideas, and new thinking do not, as the poplar adage suggests, derive from *necessity* but from playful, collective and open-ended thinking *sub specie aeterni*, not under pressure to report satisfactory financial performance within three months. Creativity and innovation do not fall from the sky, they are put to use and produced within a social setting; hence the importance of examining the dearth of instrumental rationality and short-sightedness in general. Having said that, it should be clear to the reader that this is a deeply pretentious little book, aiming to broaden the scope of economic agency and situational rationalities in organization studies, being less preoccupied with understanding, for instance, innovation work and creative activities in instrumental terms and instead recognizing that most contributions to society rely on that curious combination of hard and disciplined work as well as a playful and generous attitude. Chance indeed only favors the prepared mind, as Louis Pasteur remarked, but just how to balance chance, by definition beyond the control of us humans, and the actual preparation (i.e., organizational and managerial activities) is a widely debated question. For instance, the pharmaceutical industry, struggling to maintain its output of new and innovative therapies, has learned the hard way (i.e., through the failure of a long line of costly new drug development projects, in many cases at a late stage) that it is not possible to construct an innovation mechanism strictly based on empirically-grounded principles and historically-successful approaches; however, the need always exists for an indispensable element of creativity and chance (Styhre and Sundgren, 2011). Despite the remarkable technologies and tools being developed in, for instance, the technosciences, there is still an element of human creativity that seems to have been lost in pursuit of translating scientific expertise into commodities and capital. This calls for a critical reevaluation of the concepts of economic agency and innovation.

Chapter 1.

Innovation and the creation of economic value

Introduction

> "Rather than the highest achievement of evolution, the intellect is thus always a step behind the real, that is, the dynamic force and impetus for matter and for the creation of life that evolutionary development makes possible. Life is always richer, more complex and more simple, more diverse and more unified than the intellect can comprehend."
>
> *Elizabeth Grosz (2004: 192)*

The case of Google, the search engine and IT services company located in Mountain View, California, is one of the most well-worn examples of how innovative and creative capacities are poorly captured by economic models and accounting procedures. Nevertheless, here comes another case of the mismatch between calculated economic value – the "actual" book value – and the "virtual" value of Google:

> [A]t the start of 2009 Google was worth approximately $100 billion but had only $5 billion in physical assets and about $18 billion in cash, investments, and receivables (according to balance-sheet information and financial-market data for December 31, 2008; total financial value is the sum of market capitalization and liabilities). The other $77 billions consisted of intangible assets that the market values but which are not directly observable on the balance sheet. Because the literature is not yet well developed, we expect to see more work in this area in

the coming years. Various researchers have estimated that the annual investment in these intangibles held by US businesses is at least $1 trillion. (Brynjolfsson and Saunders, 2010: xiii–xiv)

In a society passionately preoccupied with tracing and monitoring economic value and the use of economic resources, one of the flagship corporations of the contemporary economic regimes is valued at around *four times* its book value. In addition, the neat sum of 1 trillion US dollars is invested annually in what is referred to using the handy term "intangibles". Apparently, there is a mismatch between the values being formally and officially declared and expectations regarding what cash-flows and revenues these "intangible resources" may be capable of producing in the future. Around three-quarters of Google's estimated worth of 100 billion US dollars during the period 2008–2009 is constituted by resources that are complicated to pin down in accounting figures, regardless of all the efforts made by accounting researchers and practitioners to capture these ethereal, fluid, and fundamentally slippery corporate assets (see, for instance, Mouritsen, Larsen, and Bukh, 2001). Apparently, despite all the self-assured talk about the rationality of economic procedures and routines, their transparency and communicative quality, there is a puzzling mismatch between how corporations like Google produce economic value and how the underlying resources are accounted for. The question is, then, how these three-fourths of corporate value and annual plowing down of 1 trillion US dollars transform into products and services, "value-adding activities" that reproduce this divergence between the book value and the financial value of corporations. There seems to be some kind of ghost – perhaps "spirit" or "soul," even an Aristotelian *psukhē*, might be more adequate terms – in the economic machinery; a friendly and benevolent ghost making things happen which we cannot fully understand or at least are unable to value properly.

This kinds of companies whose book values and financial values strongly diverge can be expected to grow in importance. Robertson and Swan (2004: 129) write that "[t]here is little doubt that sectors characterized by knowledge work are growing, in some cases quite rapidly". For instance, continue Robertson and Swan (2004), the science and technology sector has been growing "on average between 4% and 16% annually over the last 15 years". Barley and Kunda (2006), too, stress that neither service work nor managerial and sales work dominate the US economy, but that, primarily, "professional and technical occupations" have been growing in importance since 1950:

> Lower skilled jobs now account for 16 percent of the workforce, but service employment has grown only 4 percent since 1960. Nor does managerial and sales work account for most of the increase in white-

collar work. Today 1.5 and 4 percent more Americans work respectively as managers and salespersons than in 1950. The largest expansion in white-collar employment has occurred among professional and technical occupations. Since 1950 professional and technical employment has more than doubled, growing from 8 percent to 18 percent of the workforce. In fact, by 1991, professional and technical workers had become the largest sector surpassing even clerical workers and operatives. (Barley and Kunda, 2006: 55–56)

Brynjolfsson and Saunders (2010: 17) present some relatively new statistics, ending in 2008, demonstrating that the professional category of "finance, insurance, real estate, rental, leasing" grew from 11.4 percent of GDP in 1950 to 20.0 percent in 2008. During the same period, the professional category of "professional and business services" grew from 3.9 percent to 12.7 percent, a three-fold increase. The domain of "Wholesaling and retail", arguably representative of the emergence of a "service economy" and a "consumer society", declined during the period from 15.1 percent in 1950 to 11.9 percent in 2008. Manufacturing, accounting for 27 percent of US GNP in 1950, was less than half of that in 2008 (11.5 percent) (Brynjolfsson and Saunders, 2010: 17; Table 2.2). Based on such evidence, it is clear that manufacturing accounts for a decreasing share of US GNP, but also that there is no significant growth in the service, retail, and hospitality sectors. Growth is occurring in the field of professional work; work that is complicated to fully monitor and control as these professional workers, using Florida's (2002: 37) expression, "[c]ontrol the means of production because it is inside their heads; they *are* the means of production". Getting what is inside all those heads onto balance sheets is no trivial managerial procedure.

Regardless of these comprehensive changes, the contemporary economy is primarily understood through the rationalist lenses of economic theory and its accompanying theories of organization and management. In this corpus of work, the economy and economic agency, mostly located within organizations and companies, but also external to such principal economies entities, are commonly portrayed as the calculated use of heterogeneous resources. In addition, economic agency is widely conceptualized as the application of rationalist practices portraying economic life as a quasi-mechanical arrangement effectively making use of social resources. In opposition to such images of economic agency as exclusively embedded in a calculative and utilitarian worldview, sociologists, historians, anthropologists, and others have demonstrated that economic agency is, in many cases, determined by social and cultural conditions that extend beyond the narrow sphere of rationalist economic behavior. The present book makes a connection between economic agency and three complementary perspectives, i.e. those of playfulness, reciprocity, and squandering (the conspicuous and symbolic waste of excess resources), in terms of being three principles underlying economic agency and innovation. Rather than postulating that the *homo oeconomicus* model of economic agency prescribed by neoclassical economic theory is the only possible and legitimate image of economic agency, there are alternative models which, in various ways, contribute toward understanding how and why economic agency is performed in contemporary society.

In what follows, a relatively substantial amount of space will be dedicated to explaining the concept of neoliberalism and its social and economic consequences in terms of the financialization of the economy, including the increased emphasis on shareholder value creation as a predominant corporate governance principle in the neoliberal era. This perspective, potentially seemingly a digression from the purpose of the research monograph, is of relevance to the critique of instrumental models of agency (e.g., *homo oeconomicus*) inherent to the financialization of the economy; a model which in many ways, it is argued here, is complicated to make use of when theorizing innovation and creativity in organizations. After a review of the literature on neoliberalism and financialization, the theoretical models of agency being developed in the social sciences, and organization theory more specifically, will be discussed; in many ways, more elaborated and sensitive to actual conditions than the instrumental models of agency. Finally, the concept of innovation, theorized in analogy with an understanding of life in the life sciences, will be introduced in order to pave the way for a critique of linear and instrumental theories of innovation.

Neoliberalism is the theory, management the practice

The concept of neoliberalism

The final years of the 1970s were, suggests Harvey (2005), revolutionary ones. The 1970s had been a shaky economic period featuring two oil crises, stagflation, and general economic and social unrest. Between 1978 and 1980, a series of events took place that put neoliberal thought at the centre of political and economic power. In China, Deng Xiaoping had started to de-regulate the Chinese economy while in the US, the Head of the Federal Reserve, Paul Volcker, had started to conduct monetary policy on the basis of monetaristic economic doctrines in order to fight inflation. Politically, Margaret Thatcher was elected British Prime Minister in 1979 while in the US, Ronald Reagan took office in 1980 (Harvey, 2005). All these events helped to establish what has been called the "neoliberal state", the state serving the purpose of ensuring the free circulation of capital. Neoliberalism became the catchphrase *par préférence* of the 1980s, being put forth as an alternative to both Keynesian economic policy, by and large incapable of explaining the economic phenomenon of stagflation – simultaneously rising unemployment and inflation – and, more importantly, the welfare state *per se*. As most commentators on neoliberalism emphasize, economic inequality reached its lowest ebb in the mid 1970s, since then growing back to the levels of the 1930s.[7] Neoliberalism has served – whether intentionally or not is debatable – to restore class power, i.e., economic resources have once again been concentrated towards the hands of the few; what Leslie Sklair (2002) calls the "transnational economic class".

The concept of neoliberalism is a notoriously vague term that includes a series of references to different economists and social thinkers, schools and traditions. "Neoliberalism cannot be understood as a *singular* set of ideas and policy prescriptions, emanating from one source", argue Plehwe and Walpen (2006: 2). "It is impossible to define neoliberalism purely theoretically … neoliberalism straddles a wide range of social, political and economic phenomenon at different levels of complexity", say Saad-Filho and Johnston (2005: 1), who continue: "Neoliberalism amalgamates insight from a range of sources, including Adam Smith, neoclassical economics, the Austrian critique of Keynesianism and Soviet-style socialism, monetarism and its new classical and 'supply-side' offspring" (Saad-Filho and Johnston, 2005: 2). Wacquant (2009), too, emphasizes the fact that neoliberalism is a "hybrid term" that is at once both a "lay term" and part of the "technical terminology" of the social sciences, making its attachment to a very confined and definite meaning complicated:

---

7 Financial data and statistics offer some evidence of the effects of neoliberalism. Duménil and Lévy, 2004: 139, Table 15.6) show a sharp return to the level prevailing at the end of the 1930s as regards the portion of assets held by the richest one percent of households (approx. 40%. See also Epstein and Jayadev, 2005; Dore, 2008: 1107). Moreover, in the 1980s, the first decade of neoliberal policy, inequality decreased in one country (Italy), remained the same in eight, and increased in ten (Duménil and Lévy, 2004: 37). According to Ericson, Barry, and Doyle (2000: 554): "Neo-liberalism chooses inequality. Within a neo-liberal regime of responsible risk taking all difference, and the inequalities that result from it, is seen as a matter of choice. If one ends up poor, unemployed, and unfulfilled, it is because of poorly thought-out risk decisions." "[N]eoliberalism was chosen by economic and political elites, especially in the USA and Britain, because they believed it would best serve their interests in turbulent economic times", adds Crotty (2005: 78). In Thatcher's Britain, the number of people living in poverty grew from five million in 1979 to close to fourteen million by 1992 (Jones, 2011: 62). Between 1984 and 1989 alone, the number of homeless Britons rocketed by 38 percent (Jones, 2011: 10). In the US, during the Reagan and Bush Sr. presidencies, poverty rates grew from 11.7 percent in 1979 to 15.1 percent by 1993 (Gordon, 1996: 100).

> Neoliberalism is an elusive and contested notion, a hybrid term awkwardly suspended between the lay idiom of political debates and the technical terminology of social science, which moreover is invoked without clear referent ... Whether singular or polymorphous, evolutionary or revolutionary the prevalent conception of neoliberalism is essentially economic: it stresses an array of market friendly policies such as labor deregulation, capital mobility, privatization, a monetarist agenda of deflation and financial economy, trade liberalization, interplay competition, and the reduction of taxation and public expenditures. (Wacquant, 2009: 306)

Neoliberalism is commonly associated with individual economists such as Friedrich Hayek, Ludwig von Mises, and Milton Freeman and with the schools they represent, e.g. the Austrian School of Economics, the Department of Economics at the University of Chicago, and the monetarist doctrine developed by Friedman, German interwar Ordo-liberalism, and the Libertarianism associated with Ayn Rand and the Harvard philosopher Robert Nozick. In the neoliberal doctrine, possibly most clearly articulated by Milton Friedman, the best-known representative of neoliberalism, perhaps, besides his mentor Friedrich von Hayek, in his *Capitalism and Freedom* (1962), the economy is portrayed as preceding society. The economy is not in the service of society; on the contrary, it is economic freedom that needs to be protected from societal influence and regulation:

> Economic arrangements play a dual role in the promotion of a free society. On the one hand, freedom in economic arrangements is itself a component of freedom broadly understood, so economic freedom is an end in itself. In the second place, economic freedom is also an indispensable means towards the achievement of political goals. (Friedman, ([1962], 2002: 8)

In Friedman's view, a "free society" is a society of unregulated "competitive capitalism". "[T]he notion that political freedom was indistinguishable (or reducible to) economic freedom runs through virtually all the post-war Chicago School work", argue Davies and McGoey (2012: 70). It is little wonder, then, that proponents of neoliberal policies such as Margaret Thatcher made significant efforts to rearticulate society as a malevolent force that puts the economy at peril; not only that, Thatcher even refused to accept that society existed at all – society is a fiction and there is nothing but individuals and at best families and loosely coupled communities. The Friedman/Thatcher thesis regarding the primacy of the economy (i.e., mildly regulated competitive capitalism) is a key element of what has been called the neoliberal theology (Clarke, 2005: 51),[8] the firm belief in a neoliberal social organization that is detached from the various shortcomings and limitations of the social actors implementing neoliberal doctrines.

Although theoretically diverse, different neoliberal perspectives share a number of themes that recur in the neoliberal corpus, including the "[d]eregulation of financial markets, privatisation, weakening of institutions of social protection, weakening of labour unions and labour market protections, shrinking of gov-

8 In Madrick's (2011) account of Friedman's theories and writings, much of what Friedman proposed in terms of the primacy of economic activities consisted of unsubstantiated claims that were ignorant of actual conditions and facts. His anti-inflationary theories did not turn out to work as anticipated, leading to "slower rates of growth, higher levels of unemployment, the disappointing growth rate of productivity until the late 1990s, and stagnant wages" (Madrick, 2011: 49), not to mention what Madrick calls "extreme speculative excesses" in deregulated financial markets. In Madrick's view, Friedman's writing was always deeply ideological inasmuch as it was only marginally concerned with empirical evidence to support the various claims made: "He [Friedman] insisted he was not ideological, and adamantly claimed he based his theories on facts. In this, he exaggerated greatly. His policy essays and speeches were well written and often ingenious but overtly simple assertions of free market claims based on straightforward interpretations of Adam Smith, disregarding Smith's many caveats and philosophical and psychological writings. Friedman's social policies as opposed to his work on monetary policy were rarely substantiated by empirical research or even historical examples" (Madrick, 2011: 27. See also Kogut and Macpherson, 2011). During his entire career, Friedman was commited to a libertarian political agenda and worked actively to eliminate a number of social policies, including social security, unemployment insurance, the minimum wage, and a wide range of regulations governing labor organizing, pharmaceuticals, consumer safety and job safety. "Friedman, in effect", Madrick (2011: 27) writes, "provided the intellectual map for a reversal of the progressive evolution of the [US] nation". In neoconservatist quarters, having its center in Southern California (McGirr, 2001, 2002) and bringing first Nixon and eventually Reagan into the White House, Freidman served the role of a spokesperson for their agenda, providing seemingly solid theoretical argument against all kinds of government social programs and regulations (see also Gross, Medvetz, and Russell, 2011; Blee and Creasap, 2010). The high rates of inflation in the 1970s brought Freidman, a staunch critic of Keynesian economic theory, closely associated with the modern liberal welfare state, to the center stage in economics. Still, later statistical analyses show that Freidman's economic theory of money supply did not explain the inflation during the decade (Madrick, 2011: 48–49). In Madrick (2011) account, Freidman's only substantial and lasting constribution to economic theory is the idea that there is a minimum level of unemployment that needs to be accepted, a proposition that has been accepted also by mainstream economists. Much of the free market ideology guiding neoliberal writing and policiesduring the last decades have thus been based more on beliefs than substantial data. William K. Black (2005: 12) argues that one of the reasons why economists failed to anticipate fraud in the American savings and loan industry in the 1980s, an industry in which Black served as an regulator dring the Reagan era, was because "[p]rominent U.S. economists generally believe that regulation is the problem and deregulation is the solution". Black continues: "The deregulation ideology was the initial problem, but the fact that their [economists'] policies led to disaster also brought acute embarrassment. They had the normal human wish to avoid taking responsibility for their mistakes. Their embarrassment was particularly acute because they consider themselves the only true social scientists and believe that theory and facts, not ideology, drive their ideas." In addition, Black (2005: 13) argues, "all aspects" of the "conventional wisdom" developed by economists regarding fraud "proved false upon examination". The presence of such strong ideological beliefs prescribing deregulation stresses the "theology element" of neoliberal policy.

ernment, cutting of top tax rates, opening up of international goods and capital markets, and abandonment of full employment under the guise of the natural rate" (Pulley, 2005 25). Wacquant (2009: 307) emphasizes four elements of what he refers to as the "new institutional logic" of neoliberalism: (1) economic deregulation, (2) welfare state devolution, (3) the cultural trope of individual responsibility, and (4) an expensive, intrusive, and proactive penal apparatus. He summarizes the argument thus:

> Neoliberalism is a *transnational policies project* aiming to remake the nexus of market, state, and citizenship from above. This project is carried by a new global ruling class in the making, composed of the heads and senior executives of transnational firms, high-ranking politicians, state managers and top officials of multinational organizations (the OECD, WTO, IMF, World Bank. And the European Union), and cultural-technical experts in their employ (chief among them, economists, lawyers, and communication professionals with germane training and mental categories in their different countries). (Wacquant, 2009: 306–307)

Similarly, Griffin (2009) identifies four principles in neoliberal doctrines:

1. A confidence in the market (*marketization*) as the mechanism by which societies should be made to distribute their resources.
2. A commitment to the use of private finance (in place of public spending) in public projects (*privatization*).
3. *Deregulation*, with the removal of tariff barriers and subsidies ensuring that the market is freed from the potential tyranny of nation-state interventions, thereby providing capital with optimal mobility.
4. A commitment to *flexibilization*, which refers to the ways in which production is organized in mass consumption societies (i.e., dynamically, and flexibly) (Griffin, 2009: 9).

If it is possible to boil the neoliberal corpus down to one single, shared idea, this will be the belief in the effectiveness of market-based activities vis-à-vis governmental regulation and planning. "For neoliberalism, the market symbolises rationality in terms of an efficient distribution of resources. Government intervention, on the other hand, is deemed undesirable because it transgresses that rationality and conspires against both efficiency and liberty", argues Munck (2005: 61). Amable (2011) stresses that neoliberalism rests on two ethical principles. Firstly, the idea of *competition* is, says Amable (2011: 5), "a supreme principle" which proponents of neoliberalism believe "should be placed above political influences". Secondly, the neoliberal society institutes *self-reliance* as a social norm that guides economic policy (Amable, 2011: 6). These two ethical principles are translated into political programs and formalized economic theories. In the following, Brown (2006: 704) speaks of the "depoliticization" of capitalism *per se* in this shift from politics to market transactions: "The conversion of socially, economically, and politically produced problems into consumer items depoliticizes what has been historically produced, and it especially depoliticizes capitalism itself. Moreover, as neoliberal political rationality devolves both political problems and solutions from public to private, it further dissipates political or public life: the project of navigating the social becomes entirely one of discerning, affording, and procuring a personal solution to every socially produced problem." In other words, markets not only serve as the arena wherein economic activities are settled, they are also capable of effectively pricing and making use of resources in an optimal manner. Peet (2007: 79) speaks of the "two central principles of neoliberalism", i.e., "that 'factors of production' (labour and capital) get paid what they are worth and that free markets will not let factors go to waste". This has been called a Panglossian view of markets, i.e. that market actors are, by definition, rational in terms of valuing and pricing resources and assets effectively. A substantial body of literature criticizes neoliberal theory for overrating the effectiveness of market-based activities, commonly rejecting such forms of "market fundamentalism" (e.g., Willse, 2010; Ericson, Barry, and Doyle, 2000). Brown summarizes the shift toward neoliberal policies since the late 1970s:

> Part of what makes neoliberalism 'neo' is that it depicts free markets, free trade, and entrepreneurial rationality as achieved and normative, as promulgated through law and through social and economic policy – not simply as occurring by dint of nature. Second, neoliberalism casts the political and social spheres both appropriately dominated by market concerns and as themselves organized by the market rationality. That is, more than simply facilitating the economy, the state must construct and construe itself in market terms, as well as develop policies and promulgate a political culture that figures citizens exhaustively as rational economic actors in every sphere of life … Third, neoliberal political rationality produces governance criteria along the same lines, that is, criteria of productivity and profitability, with the consequence that governance talk increasingly becomes market-speak, business-people replace lawyers as the governing class in liberal democracies, and business principles replace juridical principles. (Brown, 2006: 694)

In the neoliberal era, juridical concerns are displaced by economic considerations, in turn shifting the focus away from politics and social planning toward consumer choice and benefit, i.e., a regime where the private is no longer political but merely a matter of making informed choices.

Neoliberalism and financialization

Tomaskovic-Devey and Lin (2011: 556) address the neoliberal policy in the US and what they regard as its single most "fundamental product", the *financialization*[9] of the economy.[10] Even though Stearns (1986) shows that the supply of investment capital grew substantially between 1949 and 1965,[11] in a historical perspective, the accumulation of capital and profit in the finance sector since the 1980s is unprecedented:

> Financial sector profits as a proportion of all profits in the economy grew slowly between 1848 and 1970, dropped across the 1970s, and increased dramatically after 1980 … This trend peaked in 2002 when 45 percent of all taxable profits in the private sector were absorbed by finance sector firms. (Tomaskovic-Devey and Lin, 2011: 539–540)

The neoclassical explanation would be, arguably, that finance sector productivity soared during the period. In a way, that is tautologically true because productivity is typically defined circulatorily in terms of "value realized in markets" (Tomaskovic-Devey and Lin, 2011: 540); "[t]he distinction between 'financial' and 'non-financial' sectors of the economy is ambiguous", says Krippner (2005: 179). The 1970s, including the oil crisis and new competition from Japan and Northern Europe in the manufacturing industry, and the resulting low-growth, high-inflation macro economy undermined the legitimacy of the "Keynesian economic solutions" predominating in the US post-World War II (Tomaskovic-Devey and Lin, 2011: 542). After a lengthy period of economic decline, starting with the first oil crisis in 1973, with lower profits and stock prices, high inflation, and unemployment, the American business community "[m]ounted a counteroffensive that

---

9 As Krippner (2005: 181) remarks, there are a number of different meanings to the term financialization. One perspective stresses shareholder value creation as the guiding principle of corporate governance; another perspective emphasizes the dominance of capital-markets over bank-based finance; a third piece of literature in the classic Marxist tradition speaks of the increased power of the *rentier* class as financialization; fourth, the concept of financialization is associated with the "explosion of financial trading associated with the proliferation of new financial instruments". Krippner (2005: 174) herself defines financialization with reference to the work of Arrighi (2010) as "a pattern of accumulation in which profits accrue primarily through financial channels rather than trade and commodity production".

10 The study by Tomaskovic-Devey and Lin (2011) addresses the case of the US, but a similar growth pattern occurred in the UK's finance sector. While the US and the UK are the leading countries in terms of the weight of their finance industries within their economies, similar tendencies are arguably visible in most other countries, while significant national and regional differences exist (see, for instance, Fiss and Zajac, 2004).

11 Between 1949 and 1965, writes Stearns (2006: 56), "gross savings increased from approximately $30 billion a year in 1945 to $150 billion a year in 1964 (or from approximately $51 billion to $160 billion in constant 1967 dollars)", consequently capital available for investment in the American economy increased by 258 percent from $511 billion to $1,827 billion (from $715 billion to $1,937 billion in constant 1967 dollars) during the period (Stearns, 2006: 59).

would dramatically remake the country's political and economic landscape" (Mizruchi and Kimeldorf, 2005: 217). This counteroffensive targeted both the labor movement and government regulations that "presumably increased the cost of doing business" (Mizruchi and Kimeldorf, 2005: 217). In this new strategy to reclaim the initiative, the Reagan presidency played a key role in serving as the vehicle for a major shift in policy:

> Reagan's election in 1980 was a turning point, particularly insofar as his vision for the country offered a more coherent ideology for the deepening assault on labor and the state. In his view, which has since become a foundation of neoliberalism, unions, regulations, or anything else that interfered with the workings of an unfettered market constituted unnecessary impediments to economic growth. By freeing up markets and implementing fiscal and tax policies designed to encourage investment, Americans, in this view, would enjoy a level of personal freedom never before experienced under the shadow of big government. (Mizruchi and Kimeldorf, 2005: 218)

"The government [the Reagan administration] … was so eager to serve the business community that firms in some cases received more than they had demanded", Mizruchi (2004: 607) argues. Reagan implemented regressive income taxes whereby the top ten percent, and especially the top one percent, would pay less payroll taxes as a percentage of their income. Unfortunately, Reagan's "supply side economics", where it was assumed that reduced taxes for the richest would "trickle down" through the economic system and lead to more jobs and boost the economy did not work as promised; the budget deficit during Reagan's first fiscal year (1982) was $130 billion (Madrick, 2011: 170–171). During his presidency, the budget deficits were in the range of 3–6.5 percent, far higher than for Carter and previous presidents. The tensions between the Federal Reserve's Paul Volcker, trying to fight inflation, and the budget deficit spending presidency of Ronald Reagan was resolved by a "massive influx of foreign investment capital, particularly from Japan", write Tomaskovic-Devey and Lin (2011: 543). This influx derived from trade surpluses in, for instance, Japan, high levels of savings in e.g., Japan and China, and an overvalued US dollar. Krippner (2011: 101–102) says:

> What the Reagan policymakers discovered in the early 1980s, then, was that they lived in a world in which capital was available in a potentially *limitless* supply. Access to global political financial markets would allow the state to defer indefinitely difficult political choices that had confronted previous administrations struggling to allocate scarce capital between competing social priorities. (Krippner, 2011: 101–102)

The foreign capital flowing into the US during the early 1980s funded the deficit and fed debt-based consumption on the part of consumers, corporations, and the US federal government alike. In addition to these macro-economic changes and new monetary policies – e.g., inflation being targeted on the basis of the monetarist doctrine that succeeded the Keynesian agenda – the financial markets were also deregulated during this period.

Since the book value of American corporations was relatively low in the late 1970s and early 1980s, the new phenomenon of "hostile takeovers" became a new concern for CEOs of major corporations (Dobbin and Zorn, 2005). Takeover specialists examined the assets of conglomerates and found that dismantling diversified major conglomerates and selling the assets individually generated substantial profits. The growth-by-diversification strategy that had dominated corporate governance in the 1970s (Davis, Diekmann, and Tinsley, 1994) thus came to an end: "In a short period of time, they [takeover specialists] gave a bad name to diversification and focused corporate attention on stock prices, because they only took over firms that were undervalued and that could be sold off, piece by piece, at a profit" (Dobbin and Zorn, 2005: 185). Theorists like Michael Jensen (see below) legitimized and justified takeover activities as a "[m]echanism for ousting poorly performing chief executives and giving control of the firms to those better suited to run them" (Dobbin and Zorn, 2005: 187). This theoretical framework, agency theory, helped takeover specialists to convince the world that what they did for a living was efficient for the economy at large and in the best interests of the stockholders. However, all other constituencies, e.g. employees and communities, traditionally part of corporate governance considerations, were regarded as irrelevant (Fligstein and Shin, 2007: 399–400). As a consequence of these new threats, CEOs became increasingly concerned about the stock market value of their firms because neglecting stock prices potentially invited hostile takeover bids which would leave them jobless (Dobbin and Zorn, 2005: 187).

The lack of regulatory oversight on a par with the inflow of capital and the growth of the finance sector, write Tomaskovic-Devey and Lin (2011: 545), encouraged "financial investment over physical capital investment and unleashed speculation in financial assets". These policies led to increased volatility in interest rates and stock market performance and, in order to counteract such uncertainty and risks (i.e., the measurable share of the total uncertainty), new financial instruments were developed that included "variable rate mortgages, credit default swaps, and mortgage-based and other derivative securities" (see also Davis, 2009a, b; Mackenzie, 2011). In addition, with the growth of institutional investors, itself an effect of the financialization of the US economy, a new active shareholder community started to influence leadership work at corporations: "The conventional wisdom ca. 1980 was that if an investor did not like the way a firm was managed, she could vote with her feet, moving her money elsewhere. Institutional investors came to believe that it made more sense to reform management than to sell off stock" (Dobbin and Zorn, 2005: 188). Institutional investors also advocated pay-for-performance schemes (i.e. performance that benefits shareholders rather than other constituencies) for executive remuneration. Eventually, new CEOs more ideologically oriented toward shareholder value schemes would displace old school CEOs, further reinforcing the shift in corporate governance. This view (see, for instance, Fligstein, 1987)[12] replaced "nonfinance sector managerial commitments to investment and innovation in specific markets", and as soon as finance-oriented managers promoted by the finance sector actors controlled major corporations, "short-term planning to increase stock prices became a primary managerial focus" (Tomaskovic-Devey and Lin, 2011: 545).

Institutionalizing agency theory

Dobbin and Zorn (2005: 181) emphasize that it was neither the traditional corporate elites nor the shareholders themselves that advocated this shift in corporate governance, rather it was "knowledge workers" in the emerging finance industry. This view is repeated in Zorn *et al.* (2005), pointing to the role of agency theory (Jensen and Meckling, 1976) as the ideological framework serving to reformulate corporate governance practices:

> Our findings suggest that the agents of change can be professional groups in financial markets who have relatively little direct contact with the corporation, but who express their preferences for firm structure and strategy through their roles in markets. And the mechanism

12 Zorn (2004) demonstrates, in contrast to Neil Fligstein's (1987, 1990) view, that finance theory trained managers gradually displaced other functional managers in a process of political games, that the use of CFOs (Chief Financial Officers) was by and large a response to new political initiatives at the end of the 1970s: "[T]he CFO solution was popularized (in no small part by finance professionals themselves) among a broad set of firms as a solution to profound regulatory change in earnings-reporting requirements after 1978" (Zorn, 2004: 349). In Zorn's (2004) view, the shift in focus from a *firm-as-a-bundle-of-productive-resources* to a *firm-as-a-bundle-of-financial-assets* was more of an unforeseen long-term consequence. Studying 429 large, public US companies over the 1963–2000 period, Zorn (2004) noticed that, while no firm had a CFO in 1963, by the end of the period, 80 percent did. The combination of high inflation at the end of the 1970s and new regulatory demands, i.e., changes in firms' environments, and responses to cope with the new situation formed the roots of the financialization of American industry rather than initially being some conscious and planned activity (see also Krippner [2011] and Abdelal [2007] for the political decisions preceding but not necessarily anticipating the financialization of the US economy). This indicates that there is, as Dobbin and Sutton (1998) suggest, a tendency among economists and management scholars to underrate the role of governments and regulations in shaping corporations: "Americans develop collective amnesia about the state's role in shaping private enterprises … Americans subscribe to the theory that firms operate in a Hobbesian economic state of nature, in which behavior depends very much on managerial initiative and markets and very little on political initiative and law" (Dobbin and Sutton, 1998: 472).

of change can be market power, which became salient to executives largely through agency theory's effect on compensation – from size-related-salaries to stock options. (Zorn et al., 2005: 287)

This new category of financial analyst was initiating a redefinition of corporate efficiency but was also the biggest beneficiary of these changes in corporate governance. Lazonick and O'Sullivan (2000) argue that the agency theory was the underlying theoretical construct justifying the rather comprehensive changes in corporate governance from the end of the 1970s:

> Given the entrenchment of incumbent corporate managers and the relatively poor performance of their companies in the 1970s, agency theorists argued that there was a need for a takeover market that, functioning as a market for corporate control, could discipline managers whose companies performed poorly. The rate of return on corporate stock was their measure of superior performance, and the maximization of shareholder value became their creed. (Lazonick and O'Sullivan, 2000: 16)

Michael C. Jensen's presidential address to the American Finance Association in Anaheim, California, in January 1993, published in the *Journal of Finance* the same year, is an exemplary text in its advocacy of an agency theory perspective in corporate governance (for a more formalist explanation of this view, see Fama and Jensen, 1983). Writing in the early 1990s, Jensen compares the 1980s with the 1890s in terms of two decades characterized by quick economic growth in the American economy. The merger boom of the 1890s, consolidating capital into larger units, led to an annual growth of total factor productivity that was almost "[s]ix times higher than that which had occurred for most of the nineteenth century" (Jensen, 1993: 834). In the 1980s, suggests Jensen (1993: 832), when "the capital markets helped eliminate excess capacity through leveraged acquisitions, stock buybacks, hostile takeovers, leveraged buyouts, and divisional sales", a similar growth in total factor productivity could be observed:

> Total factor productivity growth in the manufacturing sector more than doubled after 1981 from 1.4 percent per year in the period 1950 to 1981 to 3.3 percent in the period 1981 to 1990. Nominal unit labor costs stopped their 17-year rise, and real unit labor cost declined by 25 percent. These lower labor costs came not from reduced wages or employment but from increased productivity: Nominal and real hourly compensation increased by a total of 4.2 and 0.3 percent per year respectively over the 1981 to 1989 period. (Jensen, 1993: 836)

In addition, during the 1980s, the "real value of public firms' equity more than doubled from \$1.4 to \$3 trillion" (Jensen, 1993: 837). Jensen strongly questions the criticism of the alleged "greed" of financial actors and deplores the wide-ranging call for protectionist policies in, for instance, the manufacturing industry. Instead, Jensen advances the finance-based control of corporations as the only effective corporate governance regime. In his presidential address, Jensen is primarily concerned about market exits, the liquidation of resources in the face of insurmountable competition in certain industries, and he carefully advances a series of arguments regarding why financial analysts are in a better position to evaluate how economic resources should avoid being "wasted", i.e., unproductively invested in industries in decline. Here, Jensen stresses the inertia and myopic view of both the CEO and the board members:

> Firms often do not have good information on their own costs, much less the costs of their competitors; it is therefore sometimes unclear to managers that they are the high-cost firm which should exit the industry. Even when managers do acknowledge the requirement for exit, it is often difficult for them to accept and initiate the shutdown decision. For the managers who must implement these decisions, shutting plants or liquidating the firms causes personal pain, creates uncertainty, and interrupts or sidetracks careers. Rather than confronting the pain, managers generally resist such actions as long as they have the cash flow to subsidize the losing operations. (Jensen, 1993: 848)

In Jensen's (1993) view, there are "four control forces" influencing the corporation: (1) capital markets, (2) the legal/political/regulatory system, (3), the product and factor markets, and (4) "internal control systems headed by boards of directors" (Jensen, 1993: 850). The legal/political/regulatory system is, states Jensen (1993: 850), "far too blunt an instrument to handle the problems of wasteful managerial behavior effectively", and products and factor markets are too slow (determined by long-term contracts and other legal agreements) to effectively regulate the use of capital. In addition, there is also, continues Jensen (1993: 850), "substantial data" that supports the proposition that "the internal control systems of publicly held corporations have generally failed to cause managers to maximize efficiency and value". What remains is capital market-based control: "The capital markets [are] an effective mechanism for motivating change, renewal, and exit", says Jensen (1993: 852). To further justify a rather daring proposition like this, Jensen presents a series of statements. First, "by nature", says Jensen (1993: 852), "organizations abhor control systems, and ineffective governance is a major part of the problem with internal control mechanisms. They seldom respond in the absence of a crisis." Second, the day-to-day work of corporate boards is also ineffective in Jensen's (1993: 863) view: "Board culture is an important component of board failure. The great emphasis on politeness and courtesy at the expense of truth and frankness in boardrooms is both the symptom and cause of failure in the system."[13] The CEO is capable of controlling the flow of information to board members and to set the agenda for meetings, and this autocratic control leads to, says Jensen (1993: 864), decreased firm performance. All these declarations regarding the alleged work procedures on the executive tiers of organizations leaves the agency theorist with few opportunities other than to fully endorse capital market-based corporate governance: "The evidence from LBOs, leveraged restructurings, takeovers, and venture capital firms has demonstrated dramatically that leverage, payout policy, and ownership structure (that is, who owns the firm's securities) do in fact affect organizational efficiency, cash flow, and therefore, value", summarizes Jensen (1993: 868). Speaking in more theoretical terms, Jensen, who most conspicuously ignored most if not all of the research on accounting and corporate governance practice published by business school researchers – agency theorists, says Perrow (1986: 11), build theoretical models "almost always without empirical data" – still has the courage to call for a renewed research agenda: "Agency theory … has fundamentally changed corporate finance and organization theory, but it has yet to affect substantial research on capital-budgeting procedures. No longer can we assume managers automatically act (in opposition to their own best interests) to maximize firm value". (Jensen, 1993: 868). "We have to open up the black box called the firm", says Jensen (1993: 873), as though no researcher had ever set foot in an executive board room prior to Jensen's findings.

To summarize Jensen's arguments, the 1890s and the 1980s had the shared quality of being periods of substantial capital restructuring (through consolidation into larger units and through financialization, respectively) leading to economic growth. Jensen is concerned about the unproductive allocation of capital and the lack of market exits, i.e., the liquidation of fixed capital into financial capital. Making a series of statements regarding the nature of managerial work and the functioning of corporate boards – all fallible in Jensen's view – he deductively reaches the conclusion that financial analysts are in a better position to determine firm value and how the stock of capital should be invested. The firm antimanagerialist position of Jensen is, unfortunately, unsubstantiated as few empir-

---

13 Contrary to Jensen's statement, Davis and Robbins (2005) demonstrate, using a time series analysis of 647, 691, 691, and 822 firms during the years 1982, 1986, 1990, 1994, that corporate board composition does not affect the profitability of the firm. Davis and Robbins (2005: 291) conclude that: "[I]ndependent of performance, size, and corporate reputation, central boards are better able to attract central directors and CEOs of major corporations, leading to a relatively enduring status among corporations … Moreover, while firms that out-perform their industry are somewhat better able to recruit CEOs and central directors, there is no evidence that boards composed of these individuals enhance subsequent performance. In other words, board composition appears to be an effect of performance, not a cause."

ical studies are referenced – speaking of the firm as a "black-box" is somewhat surprising given Jensen's self-assured know-how regarding the effectiveness of internal control mechanisms – also being based on the one-dimensional view that capital-owners are the only stakeholders in firms. When Jensen speaks of "social costs" and "waste", it is not the consequences for the staff being laid off that he is concerned with, but the loss of capital investment opportunities for the capital market actors. Nowhere in his presidential address does Jensen mention the loss of job opportunities, especially stable and reasonably well-paid blue-collar work, in the American economy of the 1980s (Lazonick and O'Sullivan, 2000: 18–19). Jensen's view is therefore a strict Wall Street perspective derived from his agency theory view. Capital should be invested in the best interests of the financial market actors and, besides, little social concern needs to be shown.[14] Like a physician working at an intensive care unit, more eager to harvest the organs of an incoming patient for transplant surgery purposes than to actually save an existing patient's life (see, for instance, Lock, 2002: 101) – the patient's aggregated economic worth vis-à-vis his/her organs is higher than that of his/her life (Sharp, 2000: 296) – Jensen's agency theory is limited to the interests of a fairly small but influential group, the financial market actors; a group whose interests he defends *as though* their best interests served the interests of every stakeholder in the American economy. This is, perhaps, one of the greatest rope tricks carried out by economists; their ability to advance theories that are heavily laden with ideological assumptions as though they were disinterested fact-based theorems. Such procedures are precisely what Mackenzie and Millo (2003), as well as others, speak of as the performativity of economic theory. In many ways, Jensen's presidential address is a key document in capturing the underlying financialization ideology of the neoliberal era. Something that is somewhat surprising, however, and remains to be explained, is why remunerations paid to CEOs and corporate boards soared during the 1990s and the first decade of the new millennium (see Chapter Four) if these executives are, as Jensen repeats throughout his speech, apparently incompetent from an agency theory perspective when it comes to making use of capital resources wisely. If CEOs and board members squander capital, why should they receive a larger share of the resources they accumulate?

---

14 Mitruchi (2004) points out that Berle and Means's *The Modern Corporation & Private Property*, the seminal agency theory work, took a broader view of the agency problem than did Jensen and other more recent proponents of agency theory: "Berle and Means's concern about the separation of ownership from control was not simply about managers' lack of accountability for investors. It was also a concern about managers' lack of accountability to society in general" (Mizruchi, 2004: 581).

Consequences of shareholder value ideologies

The shift in corporate governance policy toward agency theory and shareholder value regimes was of major importance to the kinds of investments being prioritized in American industry:

> [I]nvestment in new productive capital became less attractive and financial investment became more attractive ... Institutional investors encouraged corporate CEOs to adopt the aspects of agency theory they preferred, focusing on short-term stock market value goals and tying executive compensation to stock prices. (Tomaskovic-Devey and Lin, 2011: 546)

Between 1997 and 2000, the period leading up to the first financial bubble of the new millennium, the reported profits of the S&P 500 corporations rose by 42 percent (Crotty, 2005: 102). Crotty suggests that these figures do not represent actual profits but that "financial asset prices in this era were driven by fraudulent information interpreted irrationally" (Crotty, 2005: 102). The *Wall Street Journal* addressed the shift in corporate governance priorities thus:

> These executives saw their jobs first and foremost as expanding corporate holdings, rather than managing their companies to produce better products and services. And because their focus was on immediate financial results, they also tended to see regulators as adversaries and accounting rules as inconvenient barriers to their schemes. (Wall Street Journal, June 12, 2002, "Expanding Without Managing", cited in Crotty, 2005 101)

Lazonick and O'Sullivan (2000) characterize the new regime of shareholder value corporate governance as a shift away from "from retain and reinvest" in the old regime toward "downsize and distribute" in the new: "Under the new regime, top managers downsize the corporations they control, with a particular emphasis on cutting the size of the labour forces they employ, in an attempt to increase the return on equity" (Lazonick and O'Sullivan, 2000: 18). The effects were quite substantial in terms of the loss of stable blue-collar job opportunities:[15]

---

15 As opposed to the loss of blue-collar work, and despite the fierce anti-managerialist rhetoric of agency theorists, the number of managerial positions in the US actually grew between the mid-1980s and the early 2000s: "The share of total business income devoted to managerial salaries actually rose from 16 to 23 percent between 1984 and 2001" (Goldstein, 2012: 269). When firms were downsizing, there was a need to more closely monitor the workers who remained; consequently, "strategies nominally oriented toward making firms lean and streamlined had the effect of making them fatter at the top", concludes Goldstein (2012: 277).

Hundreds of thousands of previously stable and well-paid blue-collar jobs that were lost in the recession of 1980–2 were never subsequently restored. Between 1979 and 1983, the number of people employed in the economy as a whole increased by 377,000 or 0.4 percent, while employment in durable goods manufacturing – which supplied most of the well-paid and stable blue-collar jobs – declined by 2,023,000, or 15.9 per cent. (Lazonick and O'Sullivan, 2000: 18–19)

Between 1969 and 1991, the employment rate in the fifty largest US industrial corporations in terms of sales dropped from 7 percent to 4.2 per cent of the civilian labor force (Lazonick and O'Sullivan, 2000: 19). While blue-collar work was disappearing from the American economy, the so-called pay-out ratio was increasing significantly, even during periods of profit decline:

Compared with the 1960s and 1970s, an upward shift in corporate pay-out ratios occurred in the 1980s and 1990s. In 1890, when profits declined by 17 per cent (the largest profits decline since the 1930s), dividends rose by 13 per cent, and the pay-out ratio shot up 15 points to 57 per cent. (Lazonick and O'Sullivan, 2000: 22)

In addition, argue Lazonick and Tulum (2011), the emphasis on a shareholder value ideology has led to the conscious manipulation of stock prices through stock repurchases.[16] Shareholder value ideology, write Lazonick and Tulum (2011: 1180).

[i]s an ideology that, among other things, says that any attempt by the government to interfere in the allocation of resources can only undermine economic performance. In practice, what shareholder ideology has meant for corporate resource allocation is that when companies

---

16 Corporations buy back their own stocks in cases where corporate executives believe the shares are valued too low (Lazonick, 2010: 696). In comparison to dividends, which is the owners sharing the profits generated by the company, stock buyback activities are aimed at influencing the price of the share. In Lazonick's view, dividends are associated with stability as corporations are capable of delivering profits, while stock buybacks are associated with volatility as this subjects the value of the shares to executive decisions and thus more complicated to predict. Many of the financial institutions that were part of the TARP (Troubled Assets Relief Program) bailout in 2008 in the US, had engaged in extensive stock buyback activities in the 2000–2007 period. Lazonick (2010: 696) lists the corporations and the sizes of their stock buyback programs: "Citigroup ($41.8 billion in repurchases in 2000–07), Goldman Sachs ($30.1 billion), Wells Fargo ($21.2 billion), Merrill Lynch ($21.0 billion), Morgan Stanley ($19.1 billion), American Express ($17.6 billion), and U.S. Bancorp ($12.3 billion)." Lehman Brothers and Washington Mutual, filing for bankruptcy in 2008 due to the global finance crisis, also made use of costly stock buyback programs. Companies in other sectors, too, such as pharmaceutical companies, engage in stock buyback activities rather than investing financial resources in R&D or lowering the prices of their products for the benefit of the public. "Allocated differently, the trillions spent on buybacks in the past decades could have helped stabilize the economy", contends Lazonick (2010: 696).

reap more profits they spend a substantial proportion of them on stock repurchases in an effort to boost stock prices, thus enriching first and foremost the corporate executives who make these allocative decisions. (Lazonick and Tulum, 2011: 1180)

As a consequence, the profit of the finance sector grew sharply after 1990: "After 1990, we see a second, even steeper, surge in the proportion of national profits accumulated by banks and bank holding companies" (Tomaskovic-Devey and Lin, 2011: 549). At the same time, as Fligstein and Shin (2007) show in their quantitative study of 62 US industries between 1984 and 2000, the ideology of shareholder value led to the industrial downsizing reported by Lazonick and O'Sullivan (2000), especially in industries with low profits, a lower degree of unionization, and a higher degree of computerization, but it failed, as prescribed by the underlying theoretical framework, to make these industries profitable anew: "[S]hareholder value tactics to reorganize firms and industries failed in their central goal – to create profits", contend Fligstein and Shin (2007: 401). People were thus displaced by computers and the most positive effect of the new corporate governance policy, besides the immediate benefits for finance industry actors, is, perhaps, that the computerization of American society was supported, but at the cost of lost job opportunities. Fligstein and Shin (2007) summarize their findings regarding the shift to shareholder value creation thus:

> Maximizing shareholder value and to minimize the importance of employees is a not-so-veiled way to increase profits by reducing the power of workers. Our results show that the efforts to make more profits were focused on using mergers, layoffs, and computer technologies to reorganize and remove unionized labor forces. The data suggests that workers were certainly being treated less like stakeholders and more like factors of production. (Fligstein and Shin, 2007: 420)

Goldstein, too, reporting empirical data during the period 1984–2001, stresses the assault on blue-collar jobs under the aegis of shareholder value (see also Gordon, 1996): "Ample evidence suggests that top executives learned to re-channel the anti-managerial thrust of shareholder value pressures in self-enriching ways … in part because executives' own compensation became tied to their ability to cut labor costs" (Goldstein, 2012: 269). Goldstein continues:

> Firms' attempts to stream line in accordance with shareholder value ideology differed from traditional job reductions because they did not reflect the adaptation to shifting economic circumstances so much as adaptation to an institutional environment that prescribed downsize-and-distribute as a normative orientation of corporate activity … The

reigning wisdom on Wall Street held that reducing wage rents for workers would help boost shareholder value after a decade of falling profits during the 1970s and early 1980s. Shedding unionized labor was likely an ulterior motive behind much restructuring, particularly in manufacturing industries. (Goldstein, 2012: 271–292)

Brockman, Chang, and Rennie (2007: 117) demonstrate, by studying 229 layoff announcements and the CEO remunerations paid, that CEOs engaging in downsizing are well compensated: "[T]otal compensation is 22.8% higher the year after the layoffs, and this increase also continues. These results suggest that CEOs who make layoff decisions are rewarded with higher compensation levels that persist." In addition, layoff decisions also increased shareholder value. Layoff strategies became a safe bet for CEOs wanting to both increase their individual remuneration and the market value of the firms they were directing. After the institutionalization of both agency theory and the new corporate governance regime, the finance industry could legitimately portray itself as a veritable money-machine because of its alleged ability to better allocate financial resources and spread risks (see, for instance, Ho, 2009); however, in fact, the finance sector, originally serving the so-called "real economy", did not become so much its maid as its master. The financial sector cart was put before the real economy horse. The consequences of this were that the finance sector benefitted due to its actors being able to accumulate substantial resources for themselves:

> Banking, in particular, seems to have profited most consistently from deregulation of financial markets and the resulting ability to collect economic rents from society overall. Employees of securities and commodity firms, which were historically organized as partnerships, were able to capture windfall compensation from the increasing flow of investment activity. (Tomaskovic-Devey and Lin, 2011: 549)

"Until the middle of the 20th century, money management was a craft-based vocation that did not rely on rigorous theories, analytical tools and techniques", write Lounsbury and Crumley (2007: 999). In addition, argues Stearns (2006: 48), "between the 1920s and the 1970s, financial control more or less disappeared from writings about business and the economy. Managerialism became the dominant theory of business control from the 1930s through the 1960s." Beginning in the 1970s, propelled by macroeconomic instabilities, things would change quickly. From originally being a relatively marginal domain of expertise and study at business schools (Mackenzie, 2006: 261), finance theory eventually acquired the highest status. Nobel Prizes in economics were awarded to the pioneers of finance theory, further reinforcing the institutionalization of the finance sector. Mathematically-skilled students, traditionally the recruitment base for the sciences, e.g., physics and mathematics, were attracted by a career in the finance sector because it promised the quick accumulation of wealth and was, thus, a safe bet as a human capital investment, speaking in terms of neoclassical human capital theory. Talent was directed away from the science departments toward the finance departments at elite universities, further reinforcing the tendency to privilege financial investment over other forms of capital investment as the "best and the brightest" of a generation turned their backs on the sciences, the motor of economic growth during the modern period. Taken together, all these new policies and institutional changes piped most of the profits generated in the economy controlled by the financiers into the finance sector:

> In 2008, almost a quarter of the GDP and more than a quarter of the profits accumulated in the finance sector. In the sector, employee compensation went from being above average for the economy overall to about 60 percent higher than in the rest of the economy. These shifts represent a transfer of between 5.8 and 6.6 trillion 2011 dollars in income into the finance sector, mostly as profits. (Tomaskovic-Devey and Lin, 2011: 553)

Krippner (2005) provides evidence supporting the arguments of Tomaskovic-Devey and Lin (2011):

> [M]y central empirical claim is that accumulation is now occurring increasingly through financial channels ... During 1980s and 1990s, the ratio of portfolio income to corporate cash flow ranges between approximately three and five times the levels characteristic of the 1950s and 1960s. The ratio of financial to non-financial profit behaves similarly. (Krippner, 2005: 199)

The neoclassical explanation tautologically explaining this accumulation of wealth in the finance sector, as a function of productivity growth and human capital investment, is not recognized by Tomaskovic-Devey and Lin (2011: 553):

> The finance sector's share of national income likely came at the expense of other actors in the economy. In a neoclassical economic model, one might be tempted to claim that financialization, because it may increase efficient allocation of capital, may also increase economic activity overall, thereby raising all actors' income. However, Stockhammer (2004) and Orhangazi (2008) find that financialization actually reduced nonfinancial firms' capital investment in new productive assets and increased the share of their cash flow diverted to the finance sector as increased profits. (Tomaskovic-Devey and Lin, 2011: 553)

Ultimately, the triumph of neoliberal policies and the notion of corporate governance based on shareholder value creation led to the redistribution of capital and the growth of the finance industry at the expense of other industries. In addition, short-term and linear thinking, as well as rational choice theories, became the dominant model for human agency, favoring calculative and instrumental epistemologies and accompanying ideologies.

Cultural effects of neoliberalism

In addition to the neoliberal emphasis on "capital freedom" and the absence of governmental regulation, neoliberal doctrines prescribe "individual responsibility" as their principal ethos, Ericson, Barry, and Doyle (2000: 554) say (see also Amable, 2011). Being responsible includes taking precautions to ensure one's "employability", i.e., that the individual agent acquires as much human capital to enable his/her present and future employment, to monitor and control his/her health, and to make sure that his/her income and consumption are kept in balance (Roper, Ganesh, and Inkson, 2010). For instance, concerns during recent years regarding "real estate bubbles" in the Western World (in, for instance, Ireland, Spain, and Sweden), as well as in China (e.g., in the Shanghai metropolitan area) suggest that individual households and house-owners have a problem making qualified estimations regarding how much they can invest in real estate. Since the real estate market is notoriously complicated to predict, and also a source of political regulation and financial speculation, it is hard to expect individuals to be able to make adequate judgments about their future living costs. Individual households are often in the hands of bankers when it comes to mortgages; bankers whose pay and bonuses are in many cases related to the individual agent's performance. Still, there is a belief that it is the individual house-owner who needs to be held accountable for his/her investment.

The concept of responsibility is also a significant one when studying what Fischer (2009: 14) calls "medical neoliberalism" and Pitts-Taylor (2010) refers to as "market-based health care", i.e., health care services embedded in neoliberal policy. "[M]edical neoliberalism is characterized by a commodification of health that transforms individuals from patients to consumers", argues Fischer (2009: 15), pointing to the shift in focus away from health care as the responsibility of the welfare state toward a market-based service. Pitts-Taylor (2010) and McAfee (2003) go even further, arguing that the very formulation of scientific theories draws on a neoliberal narrative, or at least that scientific theories are rephrased within neoliberal frameworks in order to make sense to the wider public. MacAfee (2003) speaks of "neoliberalism on the molecular scale" in the case of genomics and its application in order to produce genetically modified organisms

(GMOs) in agricultural research. Similarly, Pitts-Taylor (2010) suggests that findings in neurology research regarding what is called "brain plasticity" have been turned into a framework wherein intelligence and well-being are a matter self-control and training, i.e., that the functioning of the brain could be regulated on the basis of conscious choice and ambition. Pitts-Taylor (2010) writes:

> Plasticity is deployed to encourage us to see ourselves as neuronal subjects, and is linked to the continued enhancement of learning, intelligence, and mental performance, and to the avoidance of various risks associated with the brain, including mental underperformance, memory loss, and aging. While endorsing a view of the body/self which resists biological determinism, I find that the popular discourse on plasticity firmly situates the subject in a normative, neoliberal ethic of personal self-care and responsibility linked to modifying the body. (Pitts-Taylor, 2010: 639)

For Pitts-Taylor (2010: 639), "neoliberalism replaces an ethic of state care with an emphasis on individual responsibility and market fundamentalism", and this shift in focus emphasizes the individual as an enterprising, entrepreneurial agent capable of making his/her own future, even on the neurological level:

> Market-based health care policies construct populations of individuals who are encouraged to ensure their own health and promote their personal wellness and success in the face of economic insecurity and globalization; they simultaneously render patient population consumers. Health maintenance becomes a responsibility or a duty rather than a right, and bodies and selves are targeted for intense personal care and enhancement. (Pitts-Taylor, 2010: 639–670)

In the far-driven idea of market-based health care, even neurological wellness is a matter of individual ambition and self-discipline; intellectual training and daily exercise, for instance, are assumed to keep neurodegenerate diseases such as Alzheimer's and Parkinson's at bay. The responsibility of the individual extends almost infinitely.

Markets and management

In a society preoccupied with transforming welfare state activities, anchored in bureaucratic and professional routines, practices and ideologies developed during the last 150 years, into market-based products and service offerings, managerial practice has been brought to the forefront of social organization. In the welfare state, publically-employed officials, experts, and administrators handle a wide variety of social problems and opportunities; in the de-regulated economy, such endeavors are organized by private organizations and, consequently, managerial practice displaces professional and administrative routines. It is, then, no wonder that a business school education and a diploma are widely recognized as the gateway to a comfortable and reasonably predictable middle-class career and way of life. Business schools have their roots in the medieval accounting schools in Italy, the *scuola d'abbaco* where sons of merchants could learn elementary double-entry bookkeeping and calculative practices (Carruthers and Espeland, 1991: 49; Rotman, 1987). During the first half of the nineteenth century, the real century of modernization in every quarter of the Western societies, the German universities – unquestionably at the forefront of development throughout the century (Hobsbawm, 1975) – instituted chairs in what was called *Polizeiwissenschaft,* a term having little to do with the armed and authoritarian defense of society but more to do with the governance of the *polis*, the Greek term for the city state (Bynum, 1994; Foucault, 2008). The term *Polizeiwissenschaft* was not, however, exported to other countries, despite the Humboldtian University serving as the role model for the university system well into the mid-twentieth century. Instead, when the first regular business school opened at the University of Pennsylvania, the Wharton Business School, in 1881, it was the professional schools in engineering and medicine that served as the role model. Since 1881, American business schools have served as the role model for training and research in management sciences in Europe and the rest of the world. Today, more than thirteen decades after the inception of the Wharton Business School, the business school is a peculiar institution, constantly struggling to strike a balance between "practical relevance" and academic and scientific rigor (Khurana, 2007; Bennis and O'Toole, 2005). Needless to say, this puts the business school faculty in a double-bind situation whereby it is constantly accused of being neither practically relevant nor theoretically qualified. However, if business schools – or at least departments of management – have never acquired the highest academic prestige, they have won the acclaim of their students as a business school education is very popular among the young and ambitious, even to the point where, for instance, representatives of engineering schools complain that, for example, finance departments are dipping into their traditional pool of talent, i.e. mathematically gifted young men and women, leading to a brain-drain out of the engineering sciences.

The cultural significance of business school training and management, in general, in contemporary society needs to be understood within the framework of more than three decades of neoliberal deregulation. Today, schools, hospitals, and geriatric care are a few examples of activities traditionally handled by the welfare state which have been subject to corporate organization, in many cases in the ownership of venture capitalists. The companies taking over social functions are in many cases based on managerial practices rather than public service ideologies or administrative routines inherited from the early days of social modernization and the growth of the welfare state during the nineteenth and twentieth centuries. Consequently, the very idea of management and its underlying assumptions needs to be carefully examined and explored.

Economic agency and situational rationality in management thinking

Enacting rationalities

Any notion of agency is, of necessity, bound up with the predominant rationalities of the relevant period of time and its culture. Organization theorists regularly address the issue of which underlying rationalities managers and other decision-makers in organizations draw on in their day-to-day work. For instance, Gephart (1996: 95–96) says that "rationality has been the driving force of modern management", but that in the present period, there is no one single unified and universally-recognized rationality that guides social action; instead, "a cacophony of local rationalities" emerges. Thus, Gephart (1996) does not call for an abandonment of the very idea of rationality but the "decentering" of it, rendering it a source of systematic reflection and locating instrumental rationalities alongside other human faculties such as "love, hope and intuition". Many organization theorists who address rationality as an analytical term return to the very idea of instrumental rationality as what only directs a subset of the totality of human action, while at the same time being put forth as the only legitimate rationality. For instance, Grint (1997: 9) suggests that management theory, like many other forms of thought, "[t]end[s] to rationalize away the paradoxes, chance, luck, errors, subjectivities, accidents, and sheer indeterminacy of life through a prism of apparent control and rationality". Instrumental rationality, presupposing that humans are capable of executing strictly rule-governed and linear sequences of carefully-evaluated actions in most social situations, is thus more of a norm, an ideal-typical and highly moral script for action rather than an adequate description of human undertakings (March, 1991a). Townley (2008: 25) is critical of the predominance of the *homo oeconomicus* analytical model in organization theory, an embodiment of instrumental rationality, suggesting that this is a "disembedded" and "atomized" image of the human agent in terms of being "engaged in the narrow utilitarian pursuit of self-interest" and being "stripped of antecedent associations, history, and attachments". In addition, the choices and preferences of this "rational being" arise mysteriously as there is little room for social explanation of aspirations and ambitions. The human agent becomes an automata responding to various economic stimuli. Rather than being a self-regulating system of preferences and actions, human reason, suggests Townley (2008: 11), is not a "pre-given object, a pre-existent entity", instead, she continues, "reason is hewn from unreason". That is to say, the economic rationality grounded in "the means-end calculating of efficiency based on individual interest" – the rationality of *homo oeconomicus* – is complemented by a variety of alternative rationalities adhering to their own underlying logics; Townley (2008: 25) speaks of *bureaucratic rationality* and *technocratic rationality* while Starbuck (2004: 1239) addresses *scientific rationality* as a specific form of rationalitycoordinating and directing human action. Townley (2008: 132) favors the term *situational rationality* as a useful concept, defined as "the temporally and spatially located sequential and interactional rationality of daily life". Situational rationality is, thus, the operative rationality of everyday action and agency. Such operative rationalities are not, remarks Fligstein (1990:11), constructed on the basis of abstract principles, rather they are enacted within the individual's domain of work, in everyday situations:

> Actors are assumed to construct rationales for their behavior on the basis of how they view the world. Their goal and strategies result from those views and are not the product of an abstract rationality. The construction of courses of action depends greatly on the position of actors within the structure of the organization, which forms the interests and identities of actors. (Fligstein, 1990: 11)

Rather than being constructed on the basis of instrumental-technical dimensions, situational rationality helps actors to cope with specific situations and engagements. The study by Hanlon et al. (2010: 166–167) of nursing work nicely illustrates the differences between what these scholars refer to as "practical and abstract rationalities":

> [P]ractical and abstract rationalities clash within a work setting. Management desires certainty and control. It aims for a standardization and universalized knowledge because this limits the variability that stems from worker discretion and practical knowledge. Oftentimes it uses science and 'objectivity' to further these goals … This struggle results in the portrayal of abstract rationality as objective and formal rather than as the substantive rationality that it actually is. This objectivity is denied to the practical rationality of labour allowing their efforts to be characterized as subjective. (Hanlon et al., 2010: 166–167)

While instrumental and technical rationalities may be helpful in guiding the management of organizations, the everyday work of health care organizations is "[f]undamentally about interpretations and intersubjectivity", say Hanlon et al. (2010: 167) – "It is about people interpreting their bodies and then health professionals interpreting both the body and the specific needs and interpretations of the individual." Such work is not accomplished on the basis of instrumental rationalities but is, on the contrary, dependent on situational rationalities helping, for instance, practicing nurses to make adequate decisions based on their integrated professional experience (see, for instance, Benner, 1984: 42–43).

Another case of situational rationality is the practice of gambling, being persistently associated with, suggests Neal (2005), pathological behavior and psychological dysfunctions (see, for instance, Vrecko, 2008). In contrast to such a view, Neal (2005: 293) suggests that gambling is a form of joint leisure activity that undoubtedly includes calculated risk-taking but also includes a series of other positive and thrilling experiences that may be worth the cost of making strictly, statistically speaking, irrational decisions (i.e., paying a sum of money for the bet that is higher than the expected value of winning). "Gambling is not purely about making money. It is also about fun, conviviality, excitement, drama, escape … in other words, it is leisure", argues Neal (2005: 292). Thus, gambling is not effectively examined and understood on the basis of strictly economic rationalities – if a secure, maximum yield had been desired by the gambler, then corporate bonds, not horserace betting slips would be bought – but as a social practice that includes many non-economic considerations. Given such empirically-observed conditions, in the domains of both work and leisure, organization theorists are highly skeptical regarding the possibility of formulating a general definition of rationality that is detached from the domain of the situated actions constituting agency. Procedures enacted in, for instance, the domain of scientific research in a laboratory setting may be formalized into what Starbuck (2004) calls scientific rationalities, but such descriptions only capture a sub-set of the totality of the activities of the laboratory and provide a relatively schematic description of scientific procedures. Instead, situational rationalities are what newcomers to a field need to learn to master; only with time and through hard work and a reasonable amount of failure will scientists fully be able to navigate within their domain of expertise.

This shift in focus away from formal rationalities toward situational rationalities is of great importance when theorizing human agency. As agency is bound up with the formal and informal rationalities of a specific field, there is a recursive relationship between agency as a practice and rationalities as scripts for action; one side presupposes the other and the study of agency would take the analyst to the rationalities guiding and structuring the actions.

The concept of agency

The concept of agency is one of the trickiest terms in the social science vocabulary. Being intimately associated with structure, the abstract and instituted norms, beliefs, routines, identities, ideologies and so forth that regulate human behavior, agency is, for some theorists, something that emerges more or less independently of such structures, while others tend to think of agency as something that is determined by social structure. A third approach, advocated by both Pierre Bourdieu and Anthony Giddens, is to think of agency and structure as being mutually constitutive whereby a recursive relationship between agency and structure is developed; agency is enabled only with reference to socially-enacted structures but structures are reproduced by the agent's upholding of socially-legitimate beliefs and norms. Emirbayer and Mische (1998: 970, original emphasis omitted) defend such a view and speak of agency as "[t]he temporally constructed engagement by actors of different structural environments – the temporal-rational context of action – which, through the interplay of habit, imagination, and judgment, both reproduces and transforms those structures in interactive response to the problems posed by changing historical situations." Elsewhere, Emirbayer (1997) writes:

> Agency is always a dialogic process by which actors immersed in the *durée* of lived experience engage with others in collectively organized action contexts, temporal as well as spatial. Agency is path dependent as well as situationally embedded; it signifies modes of response to problems impinging upon it through sometimes broad expanses of time as well as space. (Emirbayer, 1997: 294)

Emirbayer (1997) and Emirbayer and Mische (1998) thus suggest that agency needs to be understood within the social framework wherein the agent is located; agency is "situationally embedded" and staged within the duration of the "lived experience" of the agent. However, agency is of necessity associated with some form of control and self-reflexivity, i.e., with the power to act in accordance with intensions. Sewell (1992) underlines this aspect of agency:

> To be an agent means to be capable of exerting some control over the social relations in which one is enmeshed, which in turn implies the ability to transform those social relations to some degree … agents are empowered to act with and against others by structures: they have knowledge of the schemas that inform social life and have access to some measures of human and nonhuman resources. Agency arises from the actor's knowledge of schemas, which means the ability to apply them to new contexts. Or, to put the same thing the other way around, agency arises from the actor's control of resources, which means the capacity to reinterpret or mobilize an array of resources in terms of schemas other than those that constituted the array. Agency is implied by the existence of structure. (Sewell, 1992: 20)

In Sewell's perspective, agency means the mobilization of resources, the capacity to engage a variety of social or/and material resources to hand in order to accomplish certain goals and objectives. Granovetter (1985: 487) stresses that agency is to be understood within a set of social relations which, in various ways, provide the agent with various "scripts" for legitimate actions: "Actors do not behave or decide as atoms outside of social context, nor do they adhere slavishly to a script written for them by the particular intersection of social categories that they happen to occupy. Their attempts at purposive action are instead embedded in concrete, ongoing systems of social relations."

The anthropologist Marshall Sahlins (2000) underlines the social embeddedness of agency and makes a connection between agency and the economic regime of late modern capitalism. Sahlins (2000: 181) argues that "capitalism is no sheer rationality: it is a definite form of cultural order, or a cultural order acting in a particular form". For Sahlins (2010), the capitalist economic system cannot be understood either on the basis of its religious teachings, metaphysical assumptions, or on the basis of its theoretical Überbau, but needs to be examined as the "objectification of cosmology" (Sahlins, 2010: 375), the uncoordinated behavior of human beings acting under specific interests and concerns. Sahlins (2010: 381) writes:

> [T]he economy of a society is collapsed by definition into people's economizing, which is not only an ideal form of utilitarian practice but also a specific ideal of the marketplace, not the practice of people when they come home to their families. By defining the economy as economizing, however, the economists banish the cultural schemes of persons and things that order use-value, demand and production to the unexamined limbo of what they call 'exogenous' or even 'irrational' factors. (Sahlins, 2010: 381)

Sahlins suggests that economists develop a dry and esoteric technical vocabulary that in many ways excludes and thus fails to understand the ways of "economizing" of everyday life (see, for instance, Geertz, 1978; Zelizer, 2005). "Exogenous" and "irrational factors" are thus residual rationales that cannot be accommodated by economic theory. Moe (1984), a political scientist, uses Herbert Simon's work on decision-making as an example in order to address the theoretical differences between organization theorists and economists:

> [Herbert Simon's] model of bounded rationality has been influential in arguing that empirical relevance of new elements from the psychology of decision making: e.g., memory, learning, information processing, selective attention, adaptation, socialization. Economists, given their interest in markets, generally find it easy to dismiss these sorts

of factors as *unnecessary complications*. But organization theorists, including economists now doing work on organizations, are directly concerned with individual behavior and interaction among individuals, and they often find the psychological aspects of decision making impossible to ignore. (Moe, 1984: 744. Emphasis added)

Being an anthropologist, Sahlins (2010) calls for a cultural understanding of how the material resources are both used and distributed in a specific society. Economic theory fails to understand both how economizing takes place and its cultural underpinnings; it sets the cart before the horse and mistakes the effects for the cause – cultural beliefs are what direct and coordinate human behavior, rather than economic theories, and, consequently, economic theories are relatively impotent when it comes to explaining and predicting human behavior. As Žižek (2009) has argued (with reference to Alain Badiou), the very strength of capitalism is that it is precisely *not* a distinct cultural system or civilization but, on the contrary, a set of generalized algorithms and mechanisms, for the circulation and accumulation of capital, which are possible to graft onto any specific cultural system:

> [C]apitalism is effectively not a civilization of its own, with a specific way of rendering life meaningful. Capitalism is the first socio-economic order which *de-totalizes meaning:* it is not global at the level of meaning (there is no 'global capitalist world view', no 'capitalist civilization' proper; the fundamental lesson of globalization is precisely that capitalism can accommodate itself to all civilizations, from Christian to Hindu and Buddhist). (Žižek, 2009: 25)

Therefore students of organizing capitalist economies cannot take economic theories as their starting point because such theories are unable to explain how and why economic value is created "from the bottom up". On the contrary, economic theories are in many cases so deeply ethnocentric, enacting a worldview whereby virtually all cultures, except Anglo-American capitalism, are by definition imperfect and skewed, that they fail to see any contribution made by local cultures. In other words, any deviation from the Anglo-American norm is regarded as being inherently pathological. For instance, in Weeks' (2004) review of the corporate culture literature, virtually any national culture – Scandinavian, Korean, Indian, Russian, Ibero-American, and so on – was treated and examined as something that inhibits and obstructs both the economy and organizational performance and, in cases where companies and industry performed well, this was *despite* national cultures, not thanks to them. The only legitimate corporate culture was the Anglo-American one of being individualistic, competitive, and performance-oriented; but the British and American companies were also warned about becoming *too* American. In the best of all possible worlds, this literature suggests, every single corner of the planet should thoroughly adhere to Anglo-American cultures and beliefs.

Marshal Sahlins suggests that the economy, constituted on the basis of economic agency, and social action need to be understood within a broader social and cultural context. The fields of economic sociology and economic anthropology are two sub-disciplinary fields seeking to preserve the notion that the economy is shaped and formed by cultural and social conditions. Sewell's (1992) emphasis on agency as a form of mobilizing resources has been further developed by a number of contributions. Callon (2008: 34) speaks here of "distributed action" where we no should longer talk of "individual action" but of "individual agency" as all sorts of resources are engaged in accomplishing the "distributed action": "Any action is distributed. Through it, humans as well as procedures, calculation tools, instruments, and technical devices collaborate and participate in a coordinated manner. All these entities contribute in their own way to the collective action that consequently exists", argues Callon (2008: 37). Agency is not, then, strictly located in the individual agent but emerges as the assemblage of resources being put to use. Callon (2008) refers to these assemblages as *agencement*, an idiosyncratic term first used by Gilles Deleuze to denote the heterogeneous resources constituting agency. Cooren (2006: 82) makes the distinction between agent and agency whereby the former means "what or who appears to make a difference", while the latter "simply means making a difference". Agents are, then, humans or nonhumans (actants) while agency is arguably the *agencement* making the difference. Cooren (2006: 83) continues:

> Our reconceptualization of agency helps us in the matter to the extent that different types of agencies are typically created and mobilized to fulfill organizing (to just name a few, organizational charts, contracts, ledgers, surveillance cameras, statuses, checklists, orders, memos, supervisors, supervisees, programs, procedures, etc). Like organizations, organizing can thus be understood as a hybrid phenomenon that requires the mobilization of entities with variable ontologies, which contribute to the emergence or enactment of the organizational form. (Cooren, 2006: 83)

Thus, "organizational charts, contracts, ledgers, surveillance cameras", and so forth are, for Cooren (2006), material resources that have agential powers, and influence and structure organizations. In this view, agents and agency are to be kept apart. The concept of agency thus ranges from strictly anthropocentric enactments whereby individual humans are in the position to orchestrate social changes at will – what Suchman (2007: 3) calls an "ontology of separateness"

– to more distributed theories of agency whereby tools and materiality play a key role in inducing action. Needless to say, the *homo oeconomicus* model is at the anthropocentric end of the continuum, and the distributed agency model of Callon (2008) and Cooren (2006), pushing agency to the other end where it almost dissolves into the "plenum of agencies" (Cooren, 2006: 85) being engaged. Orlikowski (2010: 128) speaks of a "relational ontology" in cases where agency is not so much attributed to individual actors or particular technologies but where "capacities for action would be studies as relational, distributed and enacted through particular instantiations" (Orlikowski, 2010: 136); "[C]apacities for action are seen to be enacted in practice and the focus is on constitutive entanglement (e.g., configurations, networks, associations, mangles, assemblages, etc.) of humans and technologies", suggests Orlikowski (2010: 135). However, on the bottom line, without refusing agency to nonhumans and material resources, it is a viable operative hypothesis to assume that humans are still the principal agent in social systems, or at least that humans are at the heart of materials and technical systems governing and regulating organizations. Consequently, agency is recognized as what emerges as recursive relations to structures as well as what draws on a variety of material resources; however, at the end of the day, it is still humans that are the foremost agents of organizations. Organizations are not machines but social systems anchored in at least some agreements regarding what to accomplish and what resources to engage in order to accomplish such goals.

Agency and institutions

Zald and Lounsbury (2010: 975) lament the fact that the discipline of organization theory has "[b]ecome a fragmented set of inward gazing communities, unnecessarily impotent as a critical voice of contemporary societal institutions and contributor to policy and public debates". Few critical accounts of social practices are articulated and the community of organization researchers is primarily preoccupied with settling internal debates and controversies. As yet another "call for relevance", Zald and Lounsbury (2010: 975) "[e]ncourage scholars to become more ambitious with their research aims, and more aggressively pursue more integrative theoretical approaches that cross narrow sub-disciplinary specialization". Elsewhere, Lounsbury (2008) has articulated a critique against neo-institutional theory in terms of failing to recognize early studies of processes of institutionalization published by, for instance, Philip Selznick, Alvin Gouldner, and Peter Blau; consequently, Lounsbury (2008: 350) says, many scholars have "[e]mployed relatively dated and caricatured versions" of institutional theory. More specifically, Lounsbury (2008: 350) suggests that neo-institutionalists embrace "the narrow, utility maximizing" rationality enacted by economics' rational choice theorists in their theoretical frameworks. As opposed to such an "economic rationality," Lounsbury (2008: 350) advocates "a broader Weberian understanding of rationality as institutionally contingent". Lounsbury (2008) says that neo-institutional theorists are primarily concerned about studying isomorphisms as they have been established, rather than offering studies of how various institutional arrangements evolve as the outcomes of social struggle and controversy. One may argue that Lounsbury (2008) is suggesting that neo-institutionalist theory is writing a kind of Whig history of institutions, wherein all instituted norms, practices, and standards are capable of being explained on the basis of some Hegelian view of history whereby what exists is rational and what is rational exists. A more adequate perspective would be to recognize that all institutions are the outcome of controversies between social actors, groups and classes and that the instituted social order is an effect of, at times, century-long debates and conflicts of interests. "Stability is often a tenuous social accomplishment that often requires ongoing negotiations and institutional work", remarks Lounsbury (2008: 357). Lounsbury (2008) thus calls for a renewal of the institutional theory research agenda:

> Institutions matter in fundamental ways and any institutional approach worth its salt should be able to assist in the understanding of questions related to the dynamics of practice … In particular, I argued that a focus on institutional rationality in the form of multiple, competing logics can be particularly fruitful. While there has been some good work in this direction, much more needs to be done to understand where logics and new practice come from and how they relate to each other. (Lounsbury, 2008: 358)

Lounsbury (2008) thus warns against institutional theorists and other social scientists uncritically adopting the economic theory model of agency, the *homo oeconomicus* ideal type; rather, they should seek to construct and enact other models of human agency. While the social sciences have offered a few alternative models on the basis of the materiality and technologies used in the process, one may seek to identify complementary literary sources advancing new images of agency and the underlying rationality.

Explaining creation and innovation

The literature on the philosophy of science in the pragmatist and constructionist tradition of thinking suggests that scientific theories are no mirrors of nature or society, instead serving to coordinate geographically disperse actors and helping to enact the object of inquiry. The epistemic objects enacted by scientific communities are in many cases fluid and fluxing and the relationship between theoretical frameworks and empirical data is in many cases non-linear and porous.

As has been suggested by, for instance, Pierre Duhem (1996) and Karl Popper (1959), theories are rarely falsified on the basis of empirical data; however, theories die hard as they are commonly accompanied by accumulated anomalies and a growing number of *ad hoc* hypotheses. Canguilhem (2008) emphasizes the fact that data (i.e., "facts", generated on the basis of a collectively-enacted theoretical framework) does not constitute theories; rather, theories live a life of their own, in many cases separated from individual elements of fact: "[T]heories are not born from the facts they coordinate, which are supposed to have given rise to them. Or, more exactly: facts give rise to theories, but they engender neither the concepts that unite them internally nor the intellectual intentions they develop" (Canguilhem, 2008: 55–56). "The ignoring of data is, in fact, the easiest and most popular mode of obtaining unity in one's thought", remarked William James (1978: 36). At the same time, this does not make all theories equally useful; there are theories that are true and useful to varying degrees but such qualities are contingent on the interests and beliefs of the actors making use of the theory:

> A theory is never true or false, but truer or falser than another, if I may say so. We consider it true from the moment when we find it hard to imagine a better theory. The criteria used to decide that one system of reason is stronger than another are drawn from a huge reservoir and vary from one question to another. (Boudon, 2003: 17)[17]

---

17 As, for instance, Lakatos (1970), Livingston (1986), Mackenzie (1999), and Epple (2011) have demonstrated, not even mathematics – for many a domain of expertise operating on a strictly logical set of axioms, propositions and theorems, that should be granted the right to use grand terms such as "truths" (Gray, 2008; Gower, 2002; Tiles, 1991; Kline, 1954) – operates on the basis of mathematical proofs separated from the production of such proofs. That is, as Livingston (1986: 14), studying mathematicians' work from an ethnomethodology perspective, says, "a proof ... is not the disengaged, material argument, but is always tied to the lived-work of that theorem's particular proving; a proof is this inextricably pairing of the proof and the associated practices of its proving." Expressed in less opaque terms, this means that even though a proof is logically correct, it will not be accepted by mathematicians until it can be presented as a sequence of operations, as a demonstration of mathematical *savoir-faire*. The validity of the proof, in brief, lies in the practice of *witnessing* the work to produce the proof (Livingston, 1986: 23):

> [T]he rigor of a mathematical proof not only lies hidden within that proof's lived-work, that a proof consists of the pair the-matermatical-proof/the-practices-of-proving-to-which-that-proof-is-irremediably-tied, and that a proof is cultivated so as to realize the material proof as a disengaged version, or account, of that proof's lived-work. In this way, a mathematical proof is itself a classical study of its own practices. Conjecturally, then, a mathematical proof is itself a classical study of mathematical theory proving, and mathematics, as a discipline, is a classical science of practical action and practical reasoning. (Livingston, 1986: 176–177)

This is the reason why mathematicians reject computer-generated proofs (MacKenzie, 1999, 1993). Mathematics does not operate on the basis of a growing corpus of unquestionable proofs but even mathematical concepts are, says MacKenzie (1999: 18), "stretchable" and consequently mathematics is a "creative process in which conjectures and proofs are subjected to criticism and improvement".

A key question in this perspective, then, is what role data, observations, and "facts" – of necessity always facts within the strict sense of the prescribed analytical framework (see Latour and Woolgar, 1979) – play in the articulation of theories? Data neither constitutes theories nor does it falsify them. Yet, it is indispensable in the production of scientific know-how – no science without data. Moreover, there is no data in the strict sense of the term without the accompanying analytical framework. It appears as if data and theory are not, contrary to official (empiricist) scientific ideologies, *compatible*.

Scholium: Life as creation

In the philosophy of life – what Ansell Pearson (1999) calls biophilosophy – the concept of life is examined as something in contrast to the living. While the living is the modalities of life, the organisms and plants immediately observable, life is *that-by-which-the living-is-living*, a force of life beyond its mere manifestations and appearances (Thacker, 2010). What Thacker (2010) calls the ontology of life was first articulated in Aristotle's *De Anima* wherein the term *psukhē* is introduced as "the principle of life", the *archē* of *zoē* (the lower form of biological life) and *bios* (human life). In *De Anima*, Aristotle thus shifts between being Aristotle-the-metaphysician, concerned with life and *psukhē*-as-principle, and Aristotle-the-biologist engaging with the living, *psukhē*-as-manifestation. "[W]hile Aristotle-the-biologist observes a set of characteristics unique to what he calls life, Aristotle-the-metaphysician struggles to articulate a coherent concept to encompass all these heterogeneous characteristics of life", argues Thacker (2010: 11). These Aristotelian questions determine the philosophy of life. Elizabeth Grosz (2011) draws a line of joint thinking between Charles Darwin – himself not a philosopher but a biologist whose work transformed Western thinking – Henri Bergson, and Gilles Deleuze inasmuch as they all are concerned with the relationship between life and matter. "For Bergson", writes Grosz (2011: 30), life is "an extension and elaboration of matter through attenuating divergence or difference". She continues:

> Life is not some mysterious alternative force, an other to matter, but the elaboration and expansion of matter, the force of concentration, winding, or folding up that matter unwinds and unfolds. Organic life is different in kind from matter, but this difference utilizes the same resources, the same forces, the same mobilities characterizing the material order. (Grosz, 2011: 31)

As such, animated matter, life, is always inventive and temporal, that is, "there can never be any real repetition but only continual invention insofar as the living carry the past along with the present" (Grosz, 2011: 31); life is *becoming*, in the continuous process of "self-differentiation", of being actualized. While life is characterized by "continual invention", there is a thin line between life and non-life; "life is always of the verge of returning to the inorganic from which its elements, its very body and energies, are drawn", says Grosz (2011: 32). While Darwin, Bergson, and Deleuze further develop the Aristotelian question of what precisely is the difference between "*psukhē*-as-principle" and "*psukhē*-as-manifestation", by developing a sophisticated philosophical vocabulary – including the terms differentiation, actualization, becoming, etc. – the caesura life-matter remains; the exact passage point between life and non-life is indeterminate.[18]

Biologists and other researchers operating in the field of the life sciences are not primarily concerned with life as an ontological principle, as *that-by-which-the living-is-living*, but preoccupied with understanding the living *per se*. Still, the question of life, how materiality begets life, haunts the life sciences; Aristotle's foundational work of splitting the metaphysical/ontological principle of life and the scientific question of the living still guides our understanding of the elementary processes of biological systems. Where does life begin and where does it end? Are viruses living beings or are they merely molecular structures interacting with living biological systems? Is brain death a death proper in comparison with other definitions? The life sciences, being perhaps the most prestigious scientific field during the contemporary period, cannot fully shrug off the question of life even though scientific ideologies prescribe the abandoning of such ontological concerns.

In the life sciences, few biologists, physicians, and biochemists believe "life is nothing but chemistry", i.e. that life is something that emerges under determinate biochemical conditions from elementary substances. At the same time, the very of idea of life is immediately present in the life sciences. Until the end of the nineteenth century, at least, the concept of vitalism was discussed in the emerging life sciences, in medicine, physiology, and the new discipline of biology. Here, vitalism denotes a "life force" innate to all living beings which cannot be explained on the basis of the biological entities involved, the tissues, organs, cells, and so forth. Vitalism has been thoroughly ridiculed for offering a kind of "ghost in the

---

18 As Helmreich (2011: 671–672) remarks, life, expressed in the prefix *bio*, is explored in a variety of technoscientific activities in industry and research settings. Reproductive technologies and terms such as biotechnology, biodiversity, bioprospecting, biosecurity, biotransfer, and molecularized biopolitics are capable of operating without a proper definition of life. Therefore, "'life' becomes a trace of the scientific and cultural practices that have asked after it, a shadow of the biological and social theories meant to capture it" (Helmreich, 2011: 674). Life is thus performatively defined, as what it is capable of accomplishing when serving as a proposition within technoscientific frameworks. Life is "for us", not "per se"; it is an effect rather than a cause. In this view, the distinction between "*psukhē*-as-principle" and "*psukhē*-as-manifestation" becomes a moot question.

machinery" argument to explain life and for pursuing "unscientific" hypotheses in its research work. While inert matter is stable and predictable, life is continuous and temporal, defying reductionist analytical methods and frameworks, i.e., procedures that overlook the fluxes and changes in all living systems. Still, the very issue of life is, by and large, excluded from scientific investigations.

The French mathematician Pierre-Simon Laplace is generally credited with excluding religious elements from the sciences, once and for all, when he replied to Napoleon that he rejected the presence of God in his system: *"Sire, je n'ai pas eu besoin de cette hypothèse"*, "Sir, I did not have any use for that hypothesis", when asked what role God played in his equations (cited by Koyré, 1959: 276). In August Comte's (1975) positivist program, articulated in the mid-nineteenth century, both theological and metaphysical explanations were excluded and the sciences were portrayed in terms of being based on formal theorizing and experimental methods and observations. Just like Laplace abandoned God as an explanatory factor so, too, have the life sciences excluded life, at least in terms of being a vitalist impulse or life force. Paradoxically, the life sciences do not seek to explain life *per se*, only the conditions *enabling* life; life is a theological or metaphysical construct that cannot be studied through the analytical methods instituted in the sciences. Consequently, the life sciences, seemingly paradoxical but a logical consequence of the professional exclusion of metaphysical speculation in scientific endeavors, *cannot explain life*. The question of what constitutes *that-by-which-the living-is-living* is abandoned.

Understanding innovation

Similar to the life scientists exploring the elementary matters of life, students of economy and economic pursuits face the problem of how vitality emerges *ex nihilo*. Life scientists seek to understand how life emerges and, similarly, economists and management researchers want to apprehend how new goods, services, and companies are produced. While life scientists explore elementary biological matter in order to understand processes and biological pathways on the cellular and molecular scale, economists and management researchers study entrepreneurship and innovation work in their pursuit of understanding how novelty and change are produced from existing resources. However, their understanding of such processes remains relatively limited.

It is beyond the scope of this book to review the literature on innovation; however, there will be a short overview of this massive corpus of texts. The classic text *par excellence* is Burns and Stalker's (1961) *The Management of Innovation*, with Israel (1992) also presenting a fine historical overview of innovation practices. Authoritative and comprehensive handbooks have been published by Fagerberg,

Mowery, and Nelson (2005) and by Van de Ven, Angle, and Poole (2000). Dodgson (2000), Utterback (1994), Slappendel (1996), and Wolf (1994) are introductory texts that provide an overview of the field. More recent concepts, or "buzzwords" such as open innovation, have been advocated by Chesbrough (2003) and subjected to critical reviews (Lichtenthaler, 2011; Huizingh, 2011; Dahlander and Gann, 2010). It is possible to categorize the innovation literature in many ways, including the industry studied, in, for instance, the pharmaceutical industry (Baba and Walsh, 2010; Furman and MacGarvie, 2009; Gassman, Reepmeyer, and von Zedtwitz, 2004), the biotechnology industry (Murray, 2002; Powell, Koput, and Smith-Doerr, 1996), the construction industry (Boland, Lyytinen, and Yoo, 2007; Barrett and Sexton, 2006; Gann, 2000), news media production (Boczkowski, 2004), the music industry (Bijsterveld and Schulp, 2004; Pinch and Trocco, 2002), the automotive industry (Callon, 1980), software engineering (Harison and Koski, 2010), financial trading (Mackenzie, 2007), professional service industries including management consulting (Semadeni and Anderson, 2010), and in universities (Grimaldi et al., 2011; Toker and Gray, 2008). Innovation studies also examine the influence and importance of a variety of organizational mechanisms, e.g. managerial control (Cardinal, 2001; Feldman, 1989), procedures for the codification of underlying knowledge and expertise (Cardinal, Alessandri, and Turner, 2001), institutional arrangements (Lounsbury and Crumley, 2007; Casper and Matraves, 2003), path-dependencies in knowledge development (Combs and Hull, 1990), organization cognition (Greve and Taylor, 2000; Nightingale, 1998), organization culture (Jassawalla and Sashittal, 2002), ideology (Miles, Snow, and Miles, 2007), organization slack (Nohria and Gulati, 1996), and corporate governance (Ramirez and Tylecore, 2004). The literature also examines how different organization forms, for example, bureaucracy (Craig, 1995; Britan, 1981), post-bureaucratic organizations (Harris, 2006), networks (Cowan and Jonard, 2009; Capaldo, 2007; Cambrosio, Keating, and Mogoutov, 2004), and virtual teams (Gibson and Gibbs, 2006) influence innovative capacities.

In a comprehensive review of the innovation literature, Crossan and Apaydin (2010: 1174) concluded that "[I]nnovation research is fragmented, poorly grounded theoretically, and not fully tested in all areas." Similarly, Pavitt (2005: 87), dedicating most of his career to studies of innovation, writes in a review chapter in an edited volume that "a growing number of 'innovation studies' show little allegiance to any particular discipline, and widely disparate theories and methods coexist in relevant journals and handbooks". Despite significant efforts to understand innovation work, there is much to learn and to know. Innovation researchers can at best offer a few general recommendations regarding the management of creativity, such as those listed by Elsbach (2009):

> (1) build organizational environments that support creative thinking
> (e.g., decentralize supervision, create cross-functional collaborations,
> open communication channels) … (2) reward behaviors known to lead
> to creative output while removing punishments for these same behav-
> iors (e.g., reward risk-taking, and learning of information outside of
> one's area of expertise, and remove punishment for failures and non-
> conformity). (Elsbach, 2009: 1043)

One of the reasons for the failure to understand how new things are constituted
on the basis of old things is that the theoretical framework employed in such
studies remains inattentive to what really makes new thinking evolve, what both
nourishes and propels creativity. In Martha Nussbaum's (2010) view, the humanities – essentially purposeful without having a clear utilitarian purpose, representing the "*Zweckmässigkeit ohne Zweck*" that Kant spoke of as being a constitutive feature of aesthetics (Holquist, 2003: 368) – train individuals to think
creatively and critically and thus precede innovation: "Innovation requires minds
that are flexible, open, and creative; literature and the arts cultivate these capacities. When they are lacking, a business culture quickly loses steam", says Nussbaum (2010: 112). In Nussbaum's view, the paradox of the humanities is that, even
though they appear to have no short-term utility, they constitute an entire intellectual culture. For instance, in Paul Leonardi's (2011) study of technological
innovation, one of the key challenges is that various professional communities
develop their own "technological concept", which in turn imposes certain cognitive limitations that both direct and blindfold the innovation work, causing the
various professional communities involved in the joint production of innovation
to fail to see how other professional communities operate on the basis of alternative technology concepts that lead to alternative problem definitions and solutions:

> Innovators are blind to the fact that others have constructed differ-
> ent problems than they have. Although they may recognize that others
> interpret a technological artefact differently than they do, they fail to
> see why they do and miss the opportunity to understand that they are
> actually in the midst of constructing very different problems and their
> identification of the technology concept is itself part of that process.
> (Leonardi, 2011: 350)

Only the intellectually-flexible and fit mind is able to see things from different
angles and to take on the role of the other. Perhaps this is what Nussbaum (2010)
suggests in her defense of the humanities in contemporary society, i.e., a regime
of knowledge promoting the capacity of individuals to transcend their own life
worlds in order to gain an elementary understanding of how other human beings

enact social realities. In many domains of innovation work, such a capacity to transcend one's own beliefs is vital when it comes to anticipating and handling, for instance, technological failures.

The field of mechanical engineering, for instance, is of necessity based on approximations from calculations and experiments which are only partially representative of actual situations and forces. "Engineering is the art of compromise" and the "art of the practical", says Petroski (1996: 3). Since engineers rarely operate on the basis of indisputable facts but on the basis of approximations, their work consists of making informed judgments. Such a constructivist view of engineering suggests that "even the most rigorous claims still contain fundamental ambiguities" (Downer, 2011: 739). That is, while the sciences construct their objects of investigation, composed of experimental systems and epistemic objects in laboratory settings, engineers operate in a "messy reality" where much, but not everything, can be known and anticipated. "Where scientists modeled simple shapes in the laboratory ... engineers had to work with complex forms ... such differences made it difficult to extrapolate 'science' to 'practice'", argues Downer (2011: 747). The skilled engineer thus knows that it is not possible to deduce the perfectly accurate design of an artifact on the basis of "models, algorithms, and blueprints" since "all tests must simplify" (Downer, 2011: 748). Expressed differently, engineers need to make "subjective judgments" about what data and evidence is relevant to their undertakings. The assessments of situations and the judgments of engineers may thus, in hindsight, prove to be wrong in, for example, the event of a disaster such as an airplane crash, or as in the much-discussed case of the NASA Challenger disaster (Feldman, 2004). Downer (2011) speaks of *epistemic accidents* in such cases; that is, "those accidents that occur because a scientific or technological assumption proves to be erroneous, even though there were reasonable and logical reasons to hold that assumption before (although not after) the event" (Downer, 2011: 748. Original emphasis omitted). Engineering is, as the civil engineer A.R. Dykes once remarked, "[t]he art of modeling materials we do not wholly understand into shapes we cannot precisely analyze so as to withstand forces we cannot properly assess, in such a way that the public has no reason to suspect the extent of our ignorance" (speech to British Institution of Structural Engineers, 1976, cited in Downer, 2011: 745). "Knowing for sure" is thus rarely, if ever, the case for engineers, but their judgment is inferred from a variety of sources jointly giving some indication of whether or not some design choices are the most adequate ones or not. Albert Camus (cited in Downer, 2011: 758) once remarked that "one cannot create experience, one must undergo it", and that predicament is definitely the case for engineered technological systems or artifacts; the engineer's know-how is truly corroborated in Popper's (1959) sense of the term as "not yet proven to be false until further notice", rather than being verified. As a consequence, engineered technological systems or artifacts, e.g. aviation systems, have been becoming increasingly reliable because technological breakdowns and other cases of malfunctioning have generated insights into, as well as understandings of, these technologies over time. Technologies can thus been seen to embody the limitations designed into them and only over time and with experience will such limitations reveal themselves. Engineering work thus always includes "unavoidable surprises" (Downer, 2011: 757) and the correction of these. Downer (2011) summarizes:

> All man-made systems, from oil refineries to financial models and pharmaceuticals, are built on theory-laden knowledge-claims that contain uncertainties and ambiguities. Constructing such artifacts requires judgments: judgments about the relevance of 'neat' laboratory knowledge to 'messy' environments, for instance, or about the similarity between tests and the phenomena they purport to represent. These judgments are usually buried deep in the systems themselves, invisible to most users and observers. Yet these are always present, and to varying degrees, they always hold the potential of 'surprise'. (Downer, 2011: 758)

Given the mismatch between tests, calculations, and experiments and actual realities, the capacity to make informed judgments is not only an engineering virtue and a professional skill but also a general prerequisite for innovation work. Since rational choice theories and the *homo oeconomicus* model poorly accommodate such capacities and skills, being at best residual to rational thinking., the "intellectual tinkering" of all innovative activities are frequently underrated and only vaguely understood, Judgment, vision, and oversight – key qualities for minds that are "flexible, open, and creative" – are valuable skills in any innovation work process, but such skills are not always examined or theorized.

In the following three chapters, three complementary principles of economic agency pertaining to creation and innovation will be examined. The first principle emphasizes the role of play and *playfulness* in all economic activities. Rather than being based on the strict application of learned rules and previously-acquired know-how, playfulness assumes the capacity to overturn the present regime of things and to shift in perspective. The second chapter of the book will examine both the "theory" of playfulness and a few applications of the concept of play in organizational settings. Second, the principle of *reciprocity* will be examined in terms of being of key relevance to understanding how social formations, of necessity, rely on the mutual exchange of data, information, and know-how but also encouragement and emotional support. That is, rather than being isolated and free-standing agents, human beings actively seek, as well as thrive on interacting with, their peers and colleagues and thus the principle of reciprocity is very central to the economic agency of human beings. Third, the principle of *squandering* prescribes that all societies have mechanisms for wasting resources – economic, intellectual, and emotional. This squandering of energy is not irrational; on the contrary, it is what makes us human; periods of accumulation and scrimping and saving are interrupted by short and extravagant periods of consumption and waste. Taken together, these three principles of economic agency offer alternative inroads into understanding how creativity is used to produce novelty and innovation and, consequently, wealth in contemporary society. Just as life does not simply derive from a combination of various chemical compounds, economic wealth does not strictly derive from administrative routines and self-interested action, but from the capacity to transcend such routines and institutionalized ways of thinking. Creativity and novelty are not, then, the outcome of procedures and activities that keep things under the strict control; on the contrary, they are what emerges either at the margins or in the midst of things, in the intersections and in the rifts and in the crevasses of everyday life, in the misunderstandings and in the failures, in the *faux pas* and even in the scandals. There is no creativity or innovation without the expenditure of energy and ideas; consequently, any theory of economic agency cannot strictly rely on instrumental reason and technical knowledge interests, but must extend into cultural and social theories of human behavior and society in order to fully recognize the underlying mechanisms of economic action.

Table 1: The three principles of economic agency.

	The principle of playfulness	*The principle of reciprocity*	*The principle of squandering*
Agency	Inventive, creative, and non-functionalist	Relational, constituted through ongoing relations.	Non-calculative, non-instrumental
Outcome	New combinations of previous know-how, new thinking, new ideas.	Socially solid and temporally extended relationships in organizational fields.	Undermines the sense of predictability and robustness of everyday existence.
Principal theorists	Johan Huizinga, development psychologists (Winnicott, Vygotsky, Piaget), Roger Caillois.	Marcel Mauss, classic anthropology (Malinowski, Radcliffe-Brown, Lévi-Strauss)	Georges Bataille
Organization practices/ implications	Innovation management, knowledge-intensive and creative industries, design and arts management	Social capital; Peer-to-peer gift economies on the Internet, Joint ventures, Mass-gifts in retailing	Organization ceremonials such as Christmas parties; excessive bonuses and pay to certain groups
Innovation practices and outcomes	Ideas and expertise are used in an open-ended and creative manner to produce innovations.	Ideas are shared and circulated across communities and organizational boundaries.	Excessive resources in organizations are used to produce novel, at times counter-intuitive thinking.
Temporality (Addressed in final chapter)	Non-linear duration and circular time.	Extended, stretching out social relations as temporal co-existence	Destructing linear time through momentary waste

Summary and conclusion

This first chapter has called attention to the dearth of economic theories assuming that the organization of the contemporary economy can be understood strictly on the basis of instrumental models on man. Contrary to economic theories of agency, three principles have been put forth in order to understand economic agency: the principle of playfulness, the principle of reciprocity, and the principle of squandering. These principles are embedded in situational rationalities inasmuch as they recognize that human beings are not socially autonomous or disembedded actors, but that they operate within a dense social and cultural setting wherein the totality of the surrounding culture influences human action. While neoliberal economic theory, e.g. finance theory and rational choice theory, have developed archetypical frameworks for human agency grounded in series of calculable choices and positions (e.g., whether to buy, hold, or sell a financial asset given certain conditions), such theories are, by and large, incapable of explaining how novelty and innovation emerge since such contributions rarely, if ever, rely on a strictly calculative mindset. Economists have thus tried to bring in additional *ad hoc* hypotheses in order to explain the dynamics of capitalism. One such *ad hoc* hypothesis is the Weberian concept of *charisma*, a "non-rational" construct suggesting that, for instance, entrepreneurs are capable of overcoming difficulties by drawing on their charms and charisma, their skill in telling convincing stories, and by creating confidence among their stakeholders. However, such *ad hoc* hypotheses only operate on the fringes of an otherwise thoroughly calculative and rational regime of thinking. Contrary to such a rationalist view of human agency, the three principles discussed in this setting emphasize that situational rationalities are not, of necessity, calculative in terms of evaluating a set of alternative choices but may be rational in recognizing a broader set of rationales, institutions, traditions, etc., i.e., the totality of beliefs and practices constituting human societies. The domain of economic agency is not, as may be suggested at times, separated from this broader sphere of human existence; on the contrary, it is located precisely in such resources. Thus, the calculative and maximizing individual serves the role of a useful heuristic in theorizing economic agency while still remaining a relatively blunt tool for understanding what Joseph Schumpeter (1942) called "the gale of creative destruction" in capitalist economic systems.

Chapter 2.

Playfulness

Exordium

When the author conducted research at a major pharmaceutical company in 2009, many of the interviewees were quite concerned about managerial control and monitoring of day-to-day activities, the strong emphasis on the quantitative reporting of, for instance, the number of molecules synthesized, and a general loss of sense as regards being part of a professional community. One of the interviewees, an analytical chemist, called for a few more joint projects defined on the basis of interests and curiosity:

> You need to have a few fun things going on and some of them don't necessarily have to 'succeed', to put it like that. You don't get the same level of creativity if everyone is thinking Lean Sigma [a management control procedure] in an organization. You'll get too little creativity and too much goal-orientation, and then these new things won't happen in the same manner. (Analytical chemist, Pharmaceutical company)

He continued:

> What advances science and development is the desire to do so, and if you lack that desire then not a lot will happen … if you enjoy what you do, then you can do it so much better … It also matters if people around you are having fun because we're social animals. (Analytical chemist, Pharmaceutical company)

In a study of a biotechnology company, a synthesis chemist underlined the fact that major innovations in the field of biochemistry and medicine are, by and large, a matter of serendipity and fortunate events whereby natural and biological systems "reveal themselves" as if by accident or chance:

> Innovation capacities lie in the capacity to make *discoveries*. And then you have this whole thing about serendipity; that's not what you're working towards. But the fact is, in the end, that's what remains. All

great innovative drugs are the outcome of serendipity. There's no human being who contrives how things should work, they've made discoveries in good [in vivo] animal models and then they've tested in all sorts of ways and then they've discovered something no one else had thought of. (Synthesis chemist, Biotech Company)

In the commercial life science work of pharmaceutical and biotechnology companies, innovation work is increasingly being translated into "innovation machines" whereby the entire process is carefully monitored and controlled, even to the point where professional scientists are calling for less rigorous control and more scope for playful and collegial interactions and exchanges. This element of play matters to organizations (Styhre, 2008).

Introduction

Leonardo went to Rome with Duke Giuliano de' Medici upon the election of Pope Leo X, who was a great student of philosophy and most especially of alchemy. In Rome, he developed a paste out of a certain type of wax and, while he walked he made inflatable animals which he blew air into, making them fly through the air; but when the air ran out, they fell to the ground. To a very strange lizard, found by the gardener of the Belvedere, he fastened some wings with a mixture of quicksilver made from scales scraped from other lizards, which quivered as it moved by crawling about. After he had fashioned eyes, a horn, and a beard for it, he tamed the lizard and kept it in a box, and all the friends to whom showed it fled in terror.

Georgio Vasari ([1550], 1991), *The Lives of the Artists*, p. 296

Vasari's *The Lives of the Artists* stands out as a seminal work in Western literature. Covering most of the key artists and architects of Renaissance Italy and Florence – the birthplace of Western modernity in Jacob Buckhardt's ([1860] 1954) authoritative statement – Vasari provided if not first-hand accounts, at least contemporary views of the greatest artists of his time. Leonardo Da Vinci is covered here in a chapter fifteen pages long, a narrative of the life, from birth to death, of a person who commonly is placed next to Newton, Descartes, and Einstein as emblematic figures of universal genius embodying the rationality of the Western cultural tradition. The polymath Leonardo was not only an artist and an architect but also one of the greatest innovators ever to have lived. Many of his ideas remained unrealizable until centuries after his death, in 1519; to date, Leonardo occupies an unrivalled position in European history as the universal genius *par préférence*. Leonardo spoke of himself, however, as an *"omo sanza lettere*", an "unlettered man", a person who had a difficult relationship with the written word. "His knowledge was without equal in all the world, but his ignorance of Latin and grammar prevented him from communicating with the learned men of his time", says Italo Calvino (1996: 78). The passage from *The Lives of the Artists* cited above tallies poorly with the image of Leonardo as a sophisticated thinker and a diligent artist serving many different masters and playing a key role in the Renaissance period. Apparently, Vasari saw value in emphasizing these playful, even puerile pleasures and joys of Leonardo. Leonardo is portrayed here as a frivolous prankster, a man of creative ideas which by no means had any utility beyond their immediate effect of causing surprise. In this account, a new image of Leonardo, rarely addressed in schoolbooks and the popular media, is presented – the playful Leonardo, amusing himself and his friends with fun and games. If nothing else, Vasari offers a charming view of what is perhaps the most outstanding figure of the early modern period, a producer of great art and sketches of many advanced machines and devices. Perhaps – we may only speculate – Vasari had more far-reaching ambitions, to point to the different sides of the great man in order to understand how creative and innovative thinking pervaded all the spheres and periods of Leonardo's life. In order to create the new, one must maintain a playful, perhaps even childish, attitude toward life.

In Friedrich Schiller's *Aesthetic Letters*, published at the end of the eighteenth century, as a series of educational letters to a Danish prince, "aesthetic training" is put forth as the principal method of cultivating moral and intellectual interest. Schiller suggests that playfulness and aesthetics are of great importance to mankind. "[T]o declare it once and for all, Man plays only when he is in the full sense of the word a man, and *he is only wholly Man when he is playing*", declares Schiller (1795/2004: 80, emphasis in the original). During the course of the modern period, the field of aesthetics has been pushed aside and is located today in the arts and, to some extent, in the engineering sciences in the form of architecture and design. Otherwise, there is little agreement with Schiller's emphasis on the aesthetic training of young people. Instead, mathematics, languages, social science and even to how to handle the new digital media have been introduced into the curriculum of young people during the contemporary period. Aesthetics largely remains a specialist field pertaining to the arts, and to a few expert professions. However, the issue of play and playfulness emphasized by Schiller is still of great importance, albeit enjoying little formal recognition. Play and playfulness are the capacity to use one's creativity in order to think in new terms, to set up new rules, and to fashion new intellectual territories previously unexplored. This playfulness is a key component of the development of stylistic competencies of firms and organizations. Without the capacity to ruthlessly overturn the present regime of things, and to recognize the need for thinking in new terms and along new routes, little new thinking will occur. As a consequence, the emergence of the post-industrial attention economy means a return to both aesthetics and play as a human capacity needing to be recognized. In this chapter, the concept of play will be examined from various perspectives in order to initially demonstrate that play and playfulness are essential to human society and then to show how such elements of play serve a role in creating sustainable competitive advantage.

The chapter is structured accordingly: First, the concept of play will be examined, especially drawing on the writings of the French social theorist Roger Caillois and his theory of play. Second, the concept of festival is discussed in terms of being the collective ritual of temporarily overturning what is taken for granted in order to be able to reproduce the predominant social order. Third, studies of play and playfulness in organizations will be examined; firstly in the form of joking and humour in organizations and then in the form of what is innate in innovation work. Finally, some concluding remarks are made regarding the importance of play and playfulness in organizations.

The elements of play in human societies

The Dutch historian Johan Huizinga coined the term *homo ludens*, "the playing man", to underline the fact that all advanced cultures and civilizations rest on their capacity to play. In Huizinga's view, all reasonably advanced forms of thinking are based on the capacity to think in new terms, to put into question and critically examine the existing order of things. These capacities to pave the way for new thinking are creative in essence and are thus based on a playful attitude whereby the familiar and taken for granted is subject to critical reflection. In Huizinga's use of it, the term *homo ludens*, then, by no means denotes any frivolous escape from responsibilities or actual conditions; rather it is an embodiment of a critical attitude and an intellectual curiosity. The British psychologist D.W. Winnicott (1971) has argued in a similar vein to Huizinga (1949) in his field of expertise, child psychology, emphasizing that creativity is a form of "life force" giving meaning to existence:

> It is creative apperception more than anything else that makes the individual feel that life is worth living. Contrasted with this is a relationship to external reality which is one of compliance, the world and its details being recognized but only as something to be fitted in with or demanding adaptation. Compliance carries with it a sense of futility for the individual and is associated with the idea that nothing matters and that life is not worth living. In a tantalizing way many individuals have experienced just enough of creative living to recognize that for most of their time they are living uncreatively, as if caught up in the creativity of someone else, or of a machine. (Winnicott, 1971: 65)

Interestingly, Winnicott (1971: 96, original in italics) located play in the intersection between the "inner psychic reality" of the child and the "external reality", suggesting that play is what connects the two in bridging the psychic reality and the material world. Similarly, the anthropologist Victor Turner (1982: 27) locates play in a stage of liminality which is, on the one hand, a domain of "sacred space-time" while on the other "subversive"; play is thus betwixt and between the regular social order and disorganized chaos, a state of inventiveness and creativity. When engaging in play, says Winnicott (1971: 110), the child establishes "an autonomous self", an active and enterprising agent capable of handling and structuring his/her external reality at will. Winnicott (1971: 57) also suggests that "playing is essentially satisfying", even when it "leads to anxiety" because play is the adventurous discovery of a life world. In Winnicott's psychology, play is perhaps the principal means allowing the child to interact with the external world, a world that needs to be examined and understood through the practice of play. Play is, thus, at the very heart of human cognition. As both Jean Piaget and Lev Vygotsky, two of the most renowned psychologists of the twentieth century, have emphasized, human intelligence is not constituted as one single unified cognitive system; rather, it is the alignment and interaction of a series of capacities and competencies that include tactile and visual perception, cognitive competencies, and linguistic skills. What separates the human infant from, for example, primates is the fact that, during the course of development, the infant is capable of combining different competencies, e.g., linguistic and narrative skills and perception. A child below the age of three, says Vygotsky (1978), is only capable of accounting for what is immediately visible in an image, but the older child can predict what is happening or about to happen in a particular scenario. In narrating forthcoming events, the older child thus bridges perceptual and cognitive and linguistic skills in a way that is unattainable for a primate. Intelligence is, thus, the capacity to bridge and combine a variety of separate but interconnected systems, while play is the single most important practice enabling the child to accomplish such alignment.

Holzner (1968), a sociologist, discusses the child's innovativeness and capacity to take alternative perspectives in play and contrasts it with the adult's regular work. Holzner (1968: 52) says that work "confronts a set of objects relevant to a task which requires that the object or their relationship be changed in some way". To be undone, further work is required and therefore "work consumes energy" and its result "cannot simply be taken back, in the way in which one can take back an erroneous thought". Work is the irreversible process of consuming energy in order to accomplish a specific end. During play, in contrast, "the player is allowed and even encouraged to adopt alternative perspectives". The attitude of the player is "detached", creating a distance between the self and the play. For instance, during children's role play, suggests Holzner (1968: 53), "the child adopts and discards different roles (and corresponding perspectives) without any irrevocable commitment". This means that the child learns to move back and forth between alternative perspectives and worldviews almost effortlessly. "Play, then, is revocable and forms a contrast with the serious single-mindedness of work; if in work the concern with reality is supreme, in play it is subordinated to the fun of sheer symbolization", suggests Holzner (1968: 53). Equally for the child and the mature player, play, says Holzner (1968: 53), "establish[es] an extensive repertory of symbolic possibilities" and consequently "makes possible the manifold of different perspectives, among which we can shift almost at will". Huizinga, a historian, Winnicott, a child psychologist, and Holzner, a sociologist, all conceive of play as encouraging intellectual curiosity and a constant shift in perspectives and worldviews. For the child, such role plays are central to his/ her socialization and his/her development of cognitive capacities, while for the adult, play offers many opportunities to move beyond the sheer utility and instrumentalism of everyday life. Especially in intellectual pursuits, a playful attitude equips the actor with the capacity to shift in terms of perspective. As a consequence, play paves the way for alternative perspectives and new cognitive routes. As will be suggested below, innovation work is both bound up with and dependent on such intellectual agility.

Social forms of play

The sociologists Norbert Elias and Eric Dunning (1986) emphasize the role of sport in what Elias has called the "civilizing process", beginning in the sixteenth century when the aristocracy and the emerging bourgeoisie began to concern themselves with manners and behaviour in the public sphere. "[S]port is closely bound up with the conditions of civilization in society at large and thus with the interplay of civilizing and de-civilizing", argue Elias and Dunning (1986: 46). In Elias and Dunning's (1986) account, sport plays a functionalist and instrumental role in civilized, modern society in terms of being a domain wherein strong emotions can be legitimately displayed. "Most human societies", propose Elias and Dunning (1986: 41), "develop some counter-measures against stress-tensions they themselves generate." In civilized society, the arenas for what Heidegger refers to using the Greek term *ek-stasis*, ecstasy, the sense of "being outside of one's self", are at least reduced, if not eliminated entirely. Differentiated societies thus provide sport as a leisure activity where "emotional arousal" is legitimately displayed:

> In advanced industrial societies, leisure activities form an enclave for
> socially approved arousal of moderate excitement behaviour in pub-

lic. One cannot understand the specific character and the specific functions which leisure has in these societies if one is not aware that, in general, the public and even the private level of emotional control has become high in comparison with that of less highly differentiated societies. (Elias and Dunning, 1986: 65)

In Elias and Dunning's (1986) view, sport is thus a key mechanism for channeling "excitement" and for reducing "stress-tensions": "In the form of leisure events … our society provides for the need to experience the upsurge of strong emotions in public – for a type of excitement which does not disturb and endanger the relative orderliness of social life as the serious type of excitement is liable to do" (Elias and Dunning, 1986: 71).

While Elias and Dunning (1986) seek to advance a sociological theory of leisure and sport, they are advancing an instrumental and functionalist theory of play that is also relatively negative in terms of conceiving of play *qua* sport as the mechanisms wherein stress, tensions, and frustration, aggregated and built up in a social regime characterized by a high degree of social control, are released. As opposed to Huizinga's more (1949) affirmative view of play, Elias and Dunning (1986) recognize few productive possibilities in play and sport; beneath the surface of the sport-lover lies a biological organism poorly equipped to operate in a society characterized by minute social control, and without the stress-tension releasing mechanisms of sport, such needs would be channelled elsewhere, potentially in less "civil" and violent ways. Unfortunately, Elias and Dunning's (1986) thesis, lending itself to empirical studies, is unaccompanied by data and evidence regarding, for instance, whether or not it is true that sport-lovers are more dissatisfied with everyday life than others, and whether or not all those expressing emotional pressures are interested in sport. In addition, Elias and Dunning (1986) downgrade the intellectual elements of sport, the appreciation of outstanding performances, and the informed evaluation of, for instance, the chances of either team winning a football game and their qualities in relation to previous seasons. Sport is, in brief, characterized by its capacity to accommodate a variety of social and human needs and is not merely a mechanism serving to "let off steam" in the advanced, differentiated society.

The French social theorist Roger Caillois, a member of the *Collège de Sociologie* in Paris during the interwar period, a loosely coupled Parisian community of social theorists and philosophers such as Georges Bataille, Michel Leiris, Pierre Klossowski, and Jean Wahl (Pearce, 2003; Richman, 2003), has examined the various forms of play in historical and contemporary societies. For both Caillois and the other members of the *Collège de Sociologie*, contemporary society is characterized by the loss of what Durkheim (1912], 1995) called the *sacred*, the sense that there are certain domains of human life that are separated from everyday life and that the boundaries between these domains and everyday life cannot be transgressed. This loss of the sacred leads to a new life situation whereby moderns have to invent new forms of transgression, in many cases the establishment of myth. For instance, in secular society, the celebrity culture is prominent evidence of the need for myths that are intelligible, yet detached from the everyday life of human beings. Hollywood celebrities, royalties, and sports stars become familiar yet isolated figures to whom the individual can relate and project his/her hopes and fears onto. In contemporary society, religious figures such as saints and prophets no longer serve to influence our lives or to guide our choices and ambitions; instead, very lively and existing, yet distanced, individuals populate our life world. The members of the *Collège de Sociologie* presented many interesting perspectives on society, not least George Bataille's theory of the general economy, discussed in the fourth chapter of this volume.

Roger Caillois is interested in understanding the role of the sacred for humans, the prohibited, serious and serene domain of human life as a contrast to the profane, everyday life existence. For Caillois (2001: 5–6), perhaps surprisingly given the role he inscribes into play, the element of play in human culture represents a form of "systematic waste": "Play is an occasion of pure waste: waste of time, energy, ingenuity, skill, and often of money for the purchase of gambling equipment." This waste, the dedication of resources to unproductive activities, is justified on the basis of the sense of mastery in play. Play "constitutes a kind of haven in which one is master of destiny. There, the player himself chooses his risks, which since they are determined in advance, cannot exceed what he has exactly agreed to put into play" (Caillois, 2001: 159). In, for instance, medieval society, human life was (using Thomas Hobbes' much repeated phrase), "poor, nasty, brutish, and short" and the medieval individual had few possibilities of protecting him or herself against disease, war, and other devastating forces beyond his/her control. Still, as Norbert Elias (1978) remarks, the medieval period was a period of relatively low social control. Here, play served the role of reducing the strains of the workday and of opening up a space wherein control could be maintained while simultaneously providing elements of chance and unpredictability that made play meaningful and exciting. "[P]lay is free activity. It is also an uncertain activity. Doubt must remain until the end, and hinges upon the denouement", says Caillois (2001: 7). For Caillois, it is important to understand the role of play in human society on the basis of the economic, cultural and material conditions prevailing there. Thus, in various ways, play signifies the underlying social order and Caillois makes the point that more advanced administrative societies such as the Roman Empire favored certain forms of play, while "pre-administrative" societies developed other forms of play. Walter Benjamin (1999: 199, D10a,

2) makes a similar observation: "As life become more subject to administrative norms, people must learn to wait more. Games of chance possess the great charm of freeing people from having to wait."

Caillois identifies four genetic types of play, named using Greek terms. First, *agôn* denotes all kinds of games that are based on any form of competition between individuals or groups of individuals. Football would, for instance, be categorized under *agôn* by Callois. Second, *alea*, named after the Greek word for dice, includes all games whereby chance plays a key role. Here, most card games and casino games would be included, sharing a quality of being based on the distribution of cards and numbers that lies beyond the control of the individual. "*Agôn* is the vindication of personal responsibility; *alea* is the negation of will, a surrender to destiny", remarks Caillois (2001: 18) when distinguishing the two forms of play. Third, Caillois speaks of *mimicry* in cases where games are based on simulation. Here, masquerades and reconstructions of historical events such as the battle of Gettysburg in "authentic uniforms" would be an example of mimicry – Uncle Toby's recreation of the siege of Namur in Lawrence Sterne's *Tristram Shandy* being one example of this kind of play. In these types of games, the participants actively engage in creating fantasies and imagery that reach beyond the everyday life world. Fourthly and finally, Caillois (2001: 12) uses the term *ilinx*, the Greek term for vertigo, for all forms of play whereby the human body is put into some kind of circulation or an unusual position. Merry-go-rounds are a classic example of *ilinx*, rotating the body around an axis and creating a sense of a loss of fixed focus. Roller-coasters are another example, a more advanced technological form of *ilinx* that creates a strong sense of being separated from the everyday life situation on *terra firma*. In addition to these four ideal-types of play, Caillois conceives of combinations of the four categories. For Caillois, play is attempted in order to establish a "perfect situation" – a situation under the full control of the individuals participating, substituting "the normal confusion of contemporary life". Play is, thus, the affirmation of agency, the capacity to act consciously and strategically under determinate conditions while an element of risk remains. Rolling the dice means a one-in-six chance of getting, say, a five, but everyday life rarely, if ever, provides such definite odds. Caillois (2001: 19) says: "In games, the role of merit or chance is clear and indisputable. It is also implied that all must play with exactly the same possibility of proving their superiority, or, on another scale, exactly the same chances of winning. In one way or another, one escapes the real world and creates another." For Caillois, then, play is a form of escape into either predictable domains that affirm one's agency or, in the case of mimicry or ilinx, a temporal "loss of the self" as one enacts alternative identities or becomes located in the situation where the regular embodiment no longer applies.

Caillois also connects forms of play with particular forms of society, suggesting that different societies favor different forms of entertainment and play:

> [P]rimitive societies, which I prefer to call 'Dionysian', be they Australian, American, or African, are societies ruled equally by masks and possessions, i.e., by *mimicry* and *ilinx*. Conversely, the Incas, Assyrians, Chinese, or Romans are orderly societies with offices, careers, codes, and ready-reckoners, with fixed and hierarchical privileges in which *agôn* and *alea*, i.e., merit and hereditary, seems to be the chief complementary elements of the games of living. In contrast to the primitive societies, these are 'rational'. In the first type there are simulation and vertigo or pantomime and ecstasy which assure the intensity and, as a consequence, the cohesion of social life. In the second type, the social nexus consists of compromise, of an implied reckoning between heredity, which is a kind of chance, and capacity, which presupposes evaluation and competition. (Caillois, 2001: 87)

No matter whether Caillois' observations are historically and anthropologically adequate or not; he still advances the intriguing thesis that different societies advance different forms of play. In addition, it may also be suggested that different social strata and gendered social groups enact their own forms of play. For instance, women's use of make-up may be regarded a form of mimicry, a fashioning of the self, a form of play that is denied men in Western mainstream societies, instituting a mild prohibition of male use of, for instance, lipstick and mascara. Instead, men can resort to forms of *alea*, which seems to be a domain dominated by men. Caillois thus paves the way for a more systematic examination of forms of games in the contemporary and historical societies, pointing to the universal need to create small social worlds and communities protected from the influence of regular society.

While the forms of play examined by Caillois (2001) are everyday life matters, more systematic forms of institutionalized and collective play also exist in society, namely those of the carnival and the festival and other events temporarily resting the continuous activities of a specific society. In the next section, the carnival and what Mikhail Bakhtin calls the carnival laughter will be examined as a form of structured play.

Carnival, festival, and social reproduction

Playfulness is, in various ways, instituted in society, both during historical and contemporary periods. Perhaps the most conspicuous form of institutionalized play is the festival or the carnival, a specific historical and cultural tradition that demands some specific attention. As historians and anthropologists emphasize, almost all "primitive" or premodern societies are organized around some annual festival of a ritual character. This ritual character derives from the elements of transgression occurring during festival, the overturning of the regular social order:

> Almost every society has festivals that have retained a ritual character over the centuries. Of particular interest to the modern inquirer are observances involving the deliberate violation of established laws; for example, celebrations in which sexual promiscuity is not only tolerated but prescribed or in which incest becomes the required practice. (Girard, 1977: 119)

Here, Girard (1977: 119) remarks that the violating of laws serves to temporarily "eliminate difference": "Family and social hierarchies are temporarily suppressed or inverted; children no longer respect their parents, servants their masters, vassals their lords." Todorov (1984: 79) suggests that such temporal eliminations of difference serve to underline the circularity of life, the continuous change that characterizes human life: "The essence of carnival lies in change, in death-birth, in destructive-creative time; carnivalesque images are basically ambivalent." Carnival and festival are thus collective events celebrating and affirming the death and rebirth of a particular social order, a recognition of the circulatory nature of life and of the seasons of the year. Briggs and Burke (2009) add that the carnivals and festivals of Europe need to be understood against the backdrop of the low levels of literacy during the period, stressing that the carnival served an educative function in describing the mechanism of society:

> The importance of public rituals in Europe, including the rituals of festivals, during the 1,000 years 500–15,000 has been explained … by the low rate of literacy at that time. What could not be recorded needed to be remembered, and what needed to be remembered had to be presented in a memorable way. Elaborate and dramatic rituals such as the coronation of kings and the homage of kneeling vassals to their seated lords demonstrated to the beholders that an important event had occurred … The word 'spectacle', commonly used in the seventeenth century, was revived in the twentieth century. (Briggs and Burke, 2009: 8–9)

In general, there are two complementary perspectives on the role of the festival. Caillois (2001b) claims that the festival is a form of collective squandering of resources, a form of sacrifice in order to reaffirm society: "It often takes several years to re-amass the amount of food and wealth ostentatiously consumed or spent, and even destroyed and wasted, for destruction and waste, as forms of excess, are at the very heart of the festival", says Caillois (2001b: 98). He continues:

> Excess constantly accompanies the festival. It is not merely epiphenomenal to the excitement it engenders. It is necessary to the success of the ceremonies that are celebrated, shares their holy quality, and like them contributes to the renewal of nature or society. In reality, this seems to be the goal of the festival. Time passes and is spent. It causes one to age and die, it is that which *wears away*. (Caillois, 2001b: 101)

In Caillois' view, the excesses of the festival are presented as a contrast to the ordeal of amassing the food and beverages consumed during short periods. Everyday life would not persist if the consumption was overtly excessive; however, without pockets of excess during the course of life, the repetitious and monotonous nature of life would be unbearable. For Caillois, then, festival is a form of the "paroxysm of society", a simultaneous act of "purification" and "renewal":

> In fact, in its purest form, the festival must be defined as the paroxysm of society, purifying and renewing it simultaneously. The paroxysm is not only its climax from a religious, but also from an economic point of view. It is the occasion for the circulation of wealth, of the most important trading, of prestige gained through the distribution of accumulated reserves. (Caillois, 2001b: 125–126)

The second perspective emphasizes not so much the squandering of resources but that the playfulness and laughter are sanctioned by the carnival. The Russian literature theorist Mikhail Bakhtin's seminal work on what he calls "carnival laughter" is a key reference here. Being a literature theorist, Bakhtin remarks that the medieval period included a series of literary genres which, in various ways, parody religious speeches and vocabularies. In that respect, the European middle ages were akin to Rome, says Bakhtin (1981: 68), in providing a variety of parodic-travestying literature: "It was Rome who taught European culture how to laugh and ridicule. But of the rich heritage of laughter that was part of the written tradition of Rome only a miniscule quantity has survived: those upon whom the transmission of this heritage depended were agelasts who elected the serious word and rejected the comic reflections as a profanation" (Bakhtin, 1981: 58–59). The term *agelast* used here deserves some clarification. It derives etymologically from Greek and means "without-laughter", "a man who does not laugh, who has no sense of humour" (Kundera, 1988: 159), and is an example of Bakhtin's often rarified vocabulary. The concept denotes ideologues who denounce laughter as a frivolous activity of human expression and reject the idea that laughter has any significant function in human societies. The agelasts are, thus, all the authorities that are in opposition to folk humor, carnival laughter, and all sources of parody and travesty. In Umberto Eco's renowned *The Name of the Rose* ([1980]1983), set during the medical period, one of the monks at the monastery is a dedicated agelast desperate not to let the documents, wherein Aristotle (the leading authority in the scholastic tradition) celebrates laughter, become widely known within the monastery; in an attempt to stop the spread of this dangerous doctrine, the actual pages of the volume are poisoned, killing an avid reader as he wets his fingers to turn to the next page. *The Name of the Rose* is, thus, staged as a struggle between the orthodox agelast, in fear of the loss of authority if even Aristotle recognizes the value of laughter, and others not sharing such concerns.

For Bakhtin, the agelasts are mistaken in their failure to understand the role of laughter during the medieval period. Bakhtin examines the literary genre of *grotesque realism*, a literary genre embodying carnival laughter. While the medieval genre of grotesque realism has a forerunner in the form of the Menippean satire, an ancient genre filled with hyperbole and exaggeration, grotesque realism still remains a specific genre on its own: "Exaggeration, hyperbolism, excessiveness are generally considered fundamental attributes of the grotesque style", says Bakhtin, 1968: 303). He continues:

> [The genre romantic grotesque] was a reaction against the cold rationalism, against official, formalistic, and logical authoritarianism; it was a rejection of that which is finished and completed, of the didactic and utilitarian spirit of the Enlighteners with their narrow and artificial optimism. (Bakhtin, 1968: 37)

The French Franciscan monk François Rabelais is the key figure in Bakhtin's analysis of grotesque realism, a writer who was popular at the time and whose popularity has survived well into the contemporary period. According to Bakhtin, Rabelais' literary style is devoid of neutrality, but actively seeks to infuse the words with emotionality:

> [Rabelais'] style is characterized by the absence of neutral words and expressions. It is colloquial speech, always addressed to somebody or talking for him, or about him. For this other party there are no neutral epithets and forms; there are either polite, laudatory, flattering, cordial words, or contemptuous, debasing, abusive ones. But even regarding the third party, there are no strictly neutral tones; neither are there any neutral words concerning objects; they are either praised or abused. (Bakhtin, 1968: 420)

In addition to the genre of grotesque realism, there were also many parodies of church rituals and speeches and, during the carnival, there were many vicious parodies of ceremonies dedicated to household animals such as the ass or the pig, rather than to regular saints. In these mock-rituals, the clergy and the church were ridiculed. For Bakhtin, laughter thus plays a key role in the maintenance of the social fabric: "Laughter … overcomes fear, for it knows no inhibitions, no limitations. Its idiom is never used by violence and authority" (Bakhtin, 1968: 90). "*Böse Menschen haben keine Lieder*", declared Bertold Brecht once. Grim people, it could also be added, are not very likely to have any jokes or laughter. For Bahktin, laughter is the human act of overturning and defamiliarizing dogmatism, a critical stance vis-à-vis social organization and order:

> Laughter purifies from dogmatism, from the intolerant and the petrified; it liberates from fanaticism and pedantry, from fear and intimidation, from dialecticism, naiveté and illusion, from the single meaning, the single level, from sentimentality. Laughter does not permit seriousness to atrophy and to be torn away from the one being, forever incomplete. It restores this ambivalent wholeness. Such is the function of laughter in the historical development of culture and literature. (Bakhtin, 1968: 123)

This specific and central, but often overlooked or ignored, role of laughter in society is associated with the institution of the carnival. Like no other social event during the year, the carnival helps – seemingly paradoxically – to establish routines for de-stabilizing society during transient periods of time and thus also helps to reproduce a particular social order: "The carnival … is a practice that both opposes and destabilizes official views of reality. In doing so, it creates itself in opposition to official culture on a series of planes, for example, from seriousness to laughter, from the dogmatic to the open, from the immutable to the contingent, and from control to identity", explains Rhodes (2001: 376). Since the carnival represents a liminal condition, a passage point between last year's and next year's society, and since it serves a dual function, to destroy and restore, the carnival is filled with ambiguities and contradictions:

> All the images of the carnival are dualistic; they unite within themselves both poles of change and crisis: birth and death (the image of pregnant death), blessing and curse (benedictory carnival curses which call simultaneously for death and birth), praise and abuse, youth and old age, top and bottom, face and backside, stupidity and wisdom. Very characteristic for carnival thinking is paired images, chosen for their contrast (high/low, fat/thin, etc.) or for their similarity (doubles/ twins). (Bakhtin, 1981: 126)

For instance, during the carnivals of Europe, it was common to crown a "carnival king", in most cases recruited from the lower strata and less respected members of society, a drunkard or the village idiot, a ritual that in a very peculiar way is a form of criticism of the political powers. Bakhtin explains:

> The primary carnivalistic act is the *mock crowning and subsequent decrowning of the carnival king* ... Crowning/decrowning is a dualistic ambivalent ritual, expressing the inevitability and at the same time the creative power of the shirt-and-renewal, the joyful relativity of all structure and order, of all authority and all (hierarchical position). Crowning already contains the idea of immanent decrowning: it is ambivalent from the very start. And he who is crowned is the antipode of the real king, a slave or a jester; this act, as it were, opens and sanctifies the inside-out world of carnival. (Bakhtin, 1981: 124)

Crowning and decrowning thus represent the cyclic nature of human life, the birth-death relationship which is simultaneously an event that provokes laughter and mockery. For Bakhtin, the carnival is the principal arena for laughter, a laughter that is ambiguous in terms of simultaneously both affirming and criticizing the existing social order. Laughter and carnival are thus two instances of human playfulness that enable the continuous accumulation of food and resources during the rest of the year. Elias Canetti (1992: 42) once remarked that the applause given by the audience as a sign of gratitude to the performers on stage represents "a brief, chaotic noise in exchange for a long, well organized one". The clapping of the audience is, although instituted as a norm, spontaneous and self-organizing, while the performance is carefully rehearsed and structured. By analogy, the carnival is the short and momentary paroxysm (in Caillois' term) of a society adhering to well-structured and carefully monitored activities during the previous year; an outbreak of spontaneity after months of ordeal in order to accumulate resources. The one activity reflects the other and the two are complementary inasmuch as the former activity needs to be assisted by the celebratory practices of acclaim.

In Bakhtin's view of carnival, it is laugher and, *ipso facto*, playfulness that serve the role of both criticizing and gratifying society. Caillois (2001b) emphasizes carnival and festival as an event of consumption and extravagance, but Bahktin (1981) stresses the intellectual critique rather than the sensual pleasures of the festival period. No matter what perspective the analyst pursues, festival and carnival are instituted social rituals where play and playfulness are put to use, to poke fun at authorities, and to ridicule stodgy and conservative religious teachings and rituals. One may say that carnival laughter represents a form of self-regulatory mechanism that helps to correct and modify unjustified social practices.

Even in authoritarian societies, power must rest on at least a minimal level of tolerance granted by the people. Under all circumstances, a society devoid of humor and laugher is a society that has lost its capacity to regulate and monitor itself. Even in totalitarian societies, jokes and laughter exist, perhaps outside of the close surveillance of the authorities, as a reminder that another world is possible.

In summary, play and playfulness may manifest themselves in the form of actual play, as examined by Caillois (2001a), but playfulness is also present in the various instituted social rituals and routines being presented as a contrast to everyday activities. These instituted rituals and routines are observable in all domains of contemporary life, from "Leisure Fridays", when relatively formally-dressed bankers or other office workers are expected to dress more casually on the last day of the working week, to lavish Christmas parties thrown by corporations in December. Play and playfulness also emerge in everyday work, especially in creative and aesthetic work, i.e., the work being brought to the forefront of the attention economy. In the next section, a few examples of how play and playfulness serve their role in organizations will be discussed.

PLAY AND PLAYFULNESS IN ORGANIZATIONS

Extravaganza and celebration in corporate rituals

The principles of playfulness suggest that organizations do not only rest on bureaucratic procedures and rule-governed practices but also on an attitude or *modus vivendi* that institutes new rules or transgresses, or even violates, preexisting ones (Sørensen and Spoelstra, 2012; Abramis, 1990). Play is not inherently in opposition to rules – on the contrary, most play demands significant amounts of rules in order to enable the freedom to play – but it presupposes a certain degree of freedom to act creatively *within* the prescribed domain of the rules. While rules are the principal mechanism of administrative endeavor, safeguarding the continuation of what is at hand, rule-breaking and rule-extension are the principal mechanisms of innovation and creativity. This locates organizations in a situation where they have to be both capable of *exploiting* the resource and know-how at hand and of *exploring* new domains of know-how and expertise (March, 1991b). Playfulness is also part of the shared rituals and ceremonials that constitute and maintain organizational cultures; these rituals are also based on idiosyncratic blends of rule-following and rule-breaking. For instance, Michael Rosen's two papers on the rituals at two New York-based firms, an annual Christmas party and a breakfast meeting, demonstrate the need for balancing rules and rule-breaking (Rosen, 1985; Rosen and Astley, 1988). In Rosen's view, these kinds of rituals are playful inasmuch as they are vaunted as joyful events when the co-workers of the two organizations are given the opportunity to enjoy the fruits of their hard work, as a form of "unproductive expenditure" to celebrate the performance, prestige, and wealth of the firms. At the same time, the co-workers are being controlled through the internalization of the hierarchical order of the organizations, carefully separating the high-status employees from the rest, and manifesting the social order of the firm in the very arrangement of the ritual of the Christmas party and the breakfast meeting. These rituals are thus the playful overturning of the capitalist ethos of "time being money" and "a saved dollar being a saved dollar", while the latent function of the ritual is to both praise productive employees and further cement hierarchical structures. In Rosen's view, the Christmas party and the breakfast meeting thus draw on the principles of playfulness (rules being enacted and temporarily overturned), reciprocity (extravagant corporate spending benefitting the individual co-workers), and squandering (hard-earned resources being given away for free) all at the same time. Playfulness, reciprocity and squandering thus become enfolded into one another in a ritual serving to reproduce organizational arrangements and professional ideologies. At the same time, if there is one single mechanism explaining the ritual of the corporate Christmas party, observable throughout the Christian West, it is the principle of playfulness, movement beyond the barren territory of bureaucratic rule-following.

Humor and joking at work

Jorge Luis Borges (2010: 157) remarks that "Bernard Shaw has stated that all intellectual labor is inherently humorous". Translating this statement into the opposite causality, Shaw would perhaps agree that humor is intellectual. At least, that is the position of Henri Bergson, strongly stressing that laughter and the comedic are bound up with intelligence and separated from emotions. "The comic, we said, appeals to the intelligence, pure and simple; laughter is incompatible with emotions" (Bergson, 1999: 126). To laugh and to see something funny. or amusing. in any event is, then, the mark of intelligence. To clarify Bergson's point, one can think of the so-called "sick joke", the joke that ceases to be funny because its topic is too emotional to bring gratification. For instance, in a racist society, it may be deemed amusing to tell stories about the stupidity of a particular ethnic group; however, as soon as racism is subject to reflection, it is no longer *comme-il-faut* to tell such jokes because the whole issue of racial and ethnic inequality becomes emotional, subject to controversy, confusion, or even shame. As a consequence, it is more commonplace to tell jokes about figures of authority than about the unfortunate members of society. When emotionality interpenetrates the telling of jokes, they fall flat. As Westwood (2007), for instance, has demonstrated, the stand-up comedian who fails to tell his/her joke and to entice the audience to laugh is often pitied, and when the stand-up comedian "bombs" (i.e., fails to make the audience laugh), the "comedy dies on the stage". What was supposed to serve as an intellectual event becomes an emotional and unnerving experience for both the performing comedian and the audience. All stand-up comedians are able to tell stories about such events, especially during the early stages of their careers. However, this Bergsonian view of humor enables us to see jokes and laughter as something that is not an entirely frivolous activity but an intellectual act or a sharing of norms and beliefs. As Umberto Eco, for instance, has noted, humor is situated and contingent to a much larger extent than, for instance, tragedy. It is much easier for us moderns to understand the despair and suffering of Medea in Euripides' play than it is to understand Aristophanes' ancient comedies because, while tragedy is embedded in the universal, comedy is bound up with the particular. Thus, the claim that there are societies without humor is an absurdity; there are, rather, forms of humor we are capable of understanding and jokes that go over our heads because they are too bound up with idiosyncratic conditions which we are poorly informed about (See Billig, 2005, for an analysis of humor as social expression). "We don't understand Chinese gestures any more than Chinese sentences", remarked Wittgenstein (1967: 40), suggesting that not only language but also entire regimes of communication are grounded in specific worldviews and cultures. The same applies to humor.

Studies of jokes and humor in organizations point to both liberating and controlling functions (Collinson, 2002; Hatch, 1997). For instance, Grugulis (2002) and Terrion and Ashforth (2002) demonstrate that new colleagues and team members control one another and structure their relationships using various forms of humor and joking. For instance, humor enables forms of criticism to be articulated, thus providing opportunities for organization members to "let off steam" without having to prepare for a full confrontation with their adversaries. A few definitions of humor have been provided in the literature. Romero and Pescosolido (2008: 397, original emphasis omitted) speak of humor as "[a]musing communications that produce positive emotions and cognitions in the individual, group or organization". Cooper (2005: 797, emphasis in the original) offers the following definition: "I define humour as *any event shared by an agent (e.g., an employee) with another individual (i.e., a target) that is intended to be amusing to the target and that the target perceives as an intentional act.*" In these two definitions, humor has a positive connotation and is "amusing". But humor can be many things and, in some cases, it imposes a sense of confusion on the actors. For instance, in the dialogs of the Marx Brothers, very much grist for the mills of humor theorists, "humour is a paradoxical form of speech and action that defeats our expectations, producing laughter with its unexpected verbal inversions, contortions, and explosions, a refusal of everyday speech that lights up the everyday", says Critchley (2007: 28). He provides a few examples of this particular and highly intellectual form of joking:

1. "Do you believe in the life to come?" "Mine was always that."

2. "Have you lived all your life in Blackpool?" "Not yet."

3. "Do me a favour and close the windows, it's cold outside" "And If I close it, will that make it warm outside?"

4. "Which of the following is the odd one out? Greed, envy, malice, anger, and kindness?" (*Pause*) "And."

5. "Gentlemen, Chicolini here may talk like an idiot, and look like an idiot, but don't let that fool you. He really is an idiot." (Cited in Critchley, 2007: 28)

The Marx Brothers here de-familiarize language and inscribe words and expressions with unusual meanings, rendering the commonsense use of language a minefield of potential misunderstandings. The question of whether someone has lived his/her entire life in a particular place is commonly understood as the life up to the present date, not the total expected lifetime; however, the everyday language is ambiguous in accommodating such sources of misunderstanding. The competent use of language, thus, rests on what Harold Garfinkel (1967) referred to as *the etcetera clause*, the additional and indispensable know-how needed to follow, for instance, written instructions. For instance, when asking if a person has lived all her life in Blackpool, the interlocutor is expected to understand whether this means "all your present life to date" or "the entire expected lifetime" without having to ask for clarification. Unless there is such a shared understanding of the everyday use of language, interlocutors will potentially end up in situations where communication is either disrupted or fails completely. However, disruptions like these are humorous, as the Marx Brothers knew very well, and very many jokes, especially on part of the eloquent Groucho, are based on this ignorance of the etcetera clause. "Are you a man or a mouse?", Groucho is provocatively asked by a lady alarmed at his lack of courage. "Put some cheese on the floor and you'll find out", replies Groucho, as though the lady were actually inquiring about what mammalian taxonomy category would apply in this case, rather than seeking to insult or challenge Groucho in terms of questioning his bravery.

In some cases, humor is used as a vehicle for addressing a rather bleak social condition that is subject to criticism. Wendt (1999) examines the Dilbert comic strips poking fun at work and managerial practice in an all-too-familiar American

corporate setting. For Wendt (1999: 2–3), this "corporate lampoonery" is based
on "fear and anxiety" and thus the smile invoked on the part of the reader is a
sad one: "When we laugh at Dilbert cartoons we are also laughing in the face
of some very complex and covert power relations. Our laughter is a reflex reaction to the irony and paradox of powerlessness and double bind situations", says
Wendt (1999: 9). Exposure to various kinds of corporate downsizing, organizational restructuring, and management fads and buzzwords more generally, being part of the everyday life world in corporate capitalism (Liu, 2004: 293), is accounted for by Dilbert in humorous terms. The life and career of organization
man is today not only fickle and fragile but also exposed to various managerial
whims. The anxieties pertaining to this existential uncertainty are effectively exploited by Dilbert.

Studies of humour within organizations show that humour is part of a communicative framework whereby difficult information and messages can be shared
using jokes and other forms of interaction based on shared humour. Bechky's
(2006) study of project work in the Hollywood film industry suggests that the
work of temporal organizations is based on strict roles and assignments and that
there is little time for the members of the project team to learn to know one another in detail and that humor is thus used to convey messages without intimidat-
ing the members of the team:

> Humor allowed for expectations and understanding of roles to be dis-
> played in a less direct, and possibly less threatening, way. It also fur-
> nished crew members with a means for the role distancing; humor
> was a safety valve that enabled them to complain about their role con-
> straints while still enacting roles appropriately and accomplishing
> their work. (Bechky, 2006: 12)

In Bechky's (2006: 14) view, from the very first moment the project co-workers
arrived on set, the project members relied on "role expectations". These role expectations are structured and modified by a combination of admonishing, thanking, and joking "[t]o accelerate familiarity with one another and reinforce role expectations" (Bechky, 2006: 15). These practices of interrelating, communicating,
and negotiating roles and performances are, says Bechky (2006: 15), "heedful" in
the sense that "crew members are paying attention to and actively encouraging
certain elements of role performances". While admonishing and thanking represent relatively straightforward modes of communication, jokes represent a more
ambiguous and subtle form of interaction and social control. Still, says Bechky
(2006: 15), "on film sets, joking strengthens the role structure, making it less
likely that role negotiations will result in great structural change". Rather than
telling a project team co-worker that he/she did not work in accordance with

expectations, jokes offer the possibility of addressing such disappointment in a more indirect and potentially less threatening way, reducing the emotional strain felt by all project team co-workers. In order to get started, and to perform at the peak of their capacities, project team co-workers in the film industry use a variety of heedful interaction strategies. Sander's (2004) study of female sex workers also emphasizes the fact that humor can be used as a mechanism for coping with emotionally-charged situations. Among female sex workers, Sanders identified at least three functions of humor:

> First, humour is a strategy of 'emotional work' employed as both a business technique and a psychological distancing strategy to manage the emotions of selling sex. Second, joking relations foster friendship, sorority and group membership. Third, jesting is a means of communicating different types of information within the group and with outsiders as well as re-interpreting life's hardship and their risk of prostitution. (Sanders, 2004: 281)

Working in a trade that is heavily gendered has a low status, and is even potentially dangerous in certain situations; sex workers need to develop certain strategies in order to individually and jointly strengthen their situation, if not practically at least emotionally. Here, joking provides a series of possibilities. In more high-brow professions, too, e.g. financial trading, various perceived emotional and cognitive difficulties are approached with a humorous attitude. Caitlin Zaloom's (2006) ethnography of derivative trading at the London Stock Exchange and the Chicago Board of Exchange sketches the image of a professional community working long hours under constant pressure to perform; consequently, self-discipline is regarded as the single most important competence of the financial trader. "[T]he question 'What makes a good trader?' yields a consistent response: discipline. Although discipline does not appear to have a place in their trading strategies, traders govern themselves with strict control", writes Zaloom (2006: 128). In Zaloom's (2006: 128) view, there are at least four "core elements of discipline":

> First, traders separate their actions on the trading floor from their lives outside; second, they control the impact of loss; third, they learn to break down the continuities between past, present, and future trades by dismantling narratives of success or failure; and fourth, they maintain acute alertness in the present moment. (Zaloom, 2006: 128)

However, to balance the discipline and all the emotional labor demanded of the financial traders, an idiosyncratic form of humour prevailed within the trader community. However, the humour used in the trading situation was not of the warm and cheerful kind, but aggressively balanced between jokes and abuse:

The humour and insults that pervade the trading floor comment on the homosocial environment and highlight man-on-man domination … But these words, even then they are used in a jocular way, maintain this aggressiveness; a slip in intonation can slide into real abuse … The cursing that colors flare-ups and everyday conversations contributes to the aesthetics of asociality. The market is a place of uninhabited action. Bodies dominate the metaphors of economic competition. Fucking and being fucked are the conventional expressions of financial dominance and ruin … The violent images of sexual domination destroy the sovereignty of the individual, subjugating his body to the will of a competitor. (Zaloom, 2006: 122–123)

As Zaloom (2006) suggests, sexual metaphor and sexual aggression dominate the metaphors being used and the traders' attempts to emotionally and cognitively handle their professional work were articulated in terms of "fucking and being fucked". Again, the line of demarcation between humor and non-humor is thin and, as soon as jokes becomes too emotional, as suggested in the Bergsonian view of laughter, they cease to be funny and turn into abuse or simply become unnerving or appalling. For instance, the sexual metaphors used are not, potentially, entirely amusing for all individuals, but may actually represent a highly gendered and rather unsophisticated trait of the profession, the cultural sense of being on top of things and having the right to act as one pleases. The use of this kind of sarcastic joke, more desperate than cheerful, was also reported by Laurie Graham (1995) in her study of the implementation of the Japanese management technique of *kaizen*, the continuous improvement of operations. For the American workers in the Japanese production unit, a so-called *transplant*, these Japanese management practices represented a rather alien regime, with many of the routines, which were often based on work team collaboration, being treated as peculiar ideas by the American workers. When the workers were expected to engage in *kaizen* work, to continually fine-tune operations and eliminate "waste" – i.e. non-value adding activities – the American workers responded to this imperative by making jokes about the "*kaizening*" of the workplace:

> Some collective resistance was expressed through jokes and humour, team members making light of company rituals and the philosophy of kaizen … kaizening, the company's philosophy of continuous improvements, was also the brunt of team jokes. When the line stopped, a team member would suggest: 'Let's kaizen that chair', or if something really went wrong, a member might say, 'I guess they kaizened that!' (Graham, 1995, p. 121)

In this case, a coping mechanism for this unfamiliar management technique was shared humor. Underlying this humorous attitude is a more deep-seated anxiety regarding the future of work, and how to respond to the expectations of the Japanese management and owners.

Korczynski's (2011) ethnographic study of the uses of humor on the shop floor of a manufacturing company, MacTell (a pseudonym) in the UK, also conceives of the uses of humor in instrumental terms as something providing workers with a repertoire of activities that can collectively help them to make sense of their work-life situation:

> Humour-use is a social process, involving not only a humour instigator but also an audience … banter and having a laugh were key lubricants underpinning the strong bonds of community that existed on the Mac-Tells shopfloor. Humour helped create community not only through the direct communication that it necessarily involved, but also through the fact that for humour to work it must be based on shared understandings. Every piece of humour that occasioned a smile or a laugh helped to cement the shared understandings of the workers in the factory. (Korczynski, 2011: 1432)

In Korczynski's (2011) view, the workers were using humor as "a key cultural resource" to avoid "the most debilitating effects of alienation". However, many of the jokes derived from the shared understanding of work at MacTell being alienating. Korczynski (2011) thus speaks of humor in terms of being a reservoir of shared know-how, of "knowing humor" rather than merely being an instant of "distracting entertainment" (Korczynski, 2011: 1434). Grinding monotony and a lack of meaning are thus functionally compensated for by means of playful joking as soon as the opportunity occurs. Similar to Bechky's (2006) film workers, Sanders (2004) sex workers, Zaloom's (2006) financial traders, and Graham's (1995) manufacturing workers, all experiencing various forms of stress and situations which alienate them (as in the case of the sex workers and the manufacturing workers) or which otherwise need some kind of emotional coping mechanism (in the film industry and the finance sector), Korczynski's (2011) manufacturing workers maintained collective joking relationships in order to create meaning in a milieu otherwise offering few such opportunities.

In other studies, humour is imposed on the workers and thus becomes a form of managerial resource or strategy. Fleming's (2005) study of the call center company VaterCorp points to the possibilities of pursuing the corporate ideology of "having fun in the workplace" (see also Fleming and Sturdy, 2011). VaterCorp was managed on the basis of what Fleming (2005) calls a "paternalist ideology" whereby the company not only provides a job and a salary but also a community of peers where there are always fun and games. At VaterCorp, you come to work but also to play and to amuse yourself, as prescribed by the corporate culture there. However, as Fleming emphasizes, these corporate and managerial ambitions and strategies did no conceal the fact that the actual call center work itself was relatively monotonous and intellectually unchallenging; Fleming suggests that the paternalist culture is an attempt to eliminate the sense of having a demeaning work life. At VaterCorp, various theme days and a very affirmative attitude toward homosexuality were two examples of the liberal and "outgoing" culture of the company. At the end of the day, concludes Fleming, the paternalist culture of VaterCorp is "very selective in what its uses and omits from the traditional suite of paternalistic motifs and practices":

> As is evident, it does not necessarily fit the ideal-type of providing job-security, housing, education outside work, insurance, crèches, etc. This is a typically casualized employment environment in which short-term contracts are the norm. Overtime pay and health benefits are non-existent for most staff. VaterCorp's paternalism appears to resemble those of independent children. It is strongly symbolic in this sense and is aimed at employee expectations and commitments (but is nevertheless connected to more concrete outcomes such as customer satisfaction, remuneration and promotion opportunities). Accordingly, employees are cast as naïve children and supervisors as paternal figures who know what is right for workers, and will protect or punish them. (Fleming, 2005: 1479)

Fleming implies that the *joie de vivre* and fun and games corporate culture of VaterCorp is a cynical attempt to conceal the loss of more substantial paternalistic concerns in terms of job security and benefits. Again, the humor and playfulness are hi-jacked by corporate policies seeking to substitute formal rights and benefits with a few workplace festivities and the conquest of cool (Frank, 1997) through a recognition of gay culture, often associated with glamour and extravaganza. Similar to the cases of Dilbert, the financial traders, and the *kaizening* of American automobile workers, the humor and playfulness at VaterCorp are potentially *mise-en-scène* under the influence of neoliberal policies and doctrines, bringing in humor in an attempt to cope with rather sad emotions concerning the lack of job security and the sense of ending up in a dead-end career (see, for instance, Warren and Fineman, 2007). Still, much humor serves the role of helping human beings to deal with the dark or depressing sides of life and great comedy is often characterized by a tint of tragedy that invokes reflection. In contemporary working life, gallows humor and sarcasm prevail because the conditions of working life offer few alternatives. In this view, humor is the intelligent way to respond to the perceived inadequacies and inconsistencies of everyday life.

Playfulness and innovation

Innovation management is a field of research where the terms play and playfulness are highly useful and relevant. In the Schumpeterian tradition of economic theory, innovation is the outcome of "combinatory capacities", i.e., the ability to bring two previously separated technologies together to create a new artefact or service. Such combinatory capacities lie at the very heart of play and playfulness, moving beyond what is immediately present and visible into the realm of the possible. Dodgson, Gann, and Salter (2005) have advocated the need to account for the concept of play in innovation management research as a key process during the early stages of the innovation process. "[T]he concept of 'play' enables the link between ideas and action. 'Play' is the medium between the 'thinking' and 'doing'", write Dodgson, Gann, and Salter (2005: 137–138). Consistent with, for instance, the work of developmental psychologists such as Lev Vygotsky, Dodgson, Gann, and Salter (2005) say that the two separated but interrelated cognitive processes of thinking and doing are integrated by playing. Play is thus the "[l]inchpin between the generation of new ideas and their articulation in practice" (Dodgson, Gann, and Salter, 2005: 138). In their glossary, Dodgson, Gann, and Salter (2005: 242) define play as: "Activities associated with the selection of new ideas to ensure they are practical, economical, targeted, and marketable, including verifying, simulating, extrapolating, interpolating, preparing, testing, validating, transforming, integrating, exploring, and prioritizing." Play is, thus, a key process in the advancement of new thinking. Prior to the work of Dodgson, Gann, and Salter (2005), Anderson (1994) made a similar argument in the innovation management literature, claiming that play is a useful term in capturing a series of skills and competencies of relevance to innovation work. Anderson (1994) writes:

> Work wears us out, even before we do it. Play energizes us, even after we're done. Play also gives direction and focus on our activities. In class the mind easily wanders. On the ball field, at the mall, or cruising singles bars, the mind is incredibly focused.

Echoing Frederick W Taylor's old reflection on why men at work act differently and more conservatively than men at play, Anderson (1994: 81) suggests that "[t]he process of play gives us energy, focus, and creativity." Dougherty and Takacs (2004) also address the concept of play in their analysis of innovative firms. Innovative firms are, they suggest, better at forming "multi-functional teams" and exploiting diverse competencies and sources of know-how. These multi-functional teams in turn rely on the joint, heedful interrelating of the members of the team, and the capacity to communicate and respect the competence and perspectives of other team members. This heedful interrelating, first coined by Weick and Roberts (1993), is in turn embedded in a playful attitude among team members; the two terms are mutually constitutive:

> Play generates heedful interrelating, while heedful interrelating struc-
> tures play: play and heedful interrelating complement each other to
> produce a sensible, manageable approach to structuring the activities
> of innovation so that streams of new products are possible, even in
> established organizations with mature technologies. (Dougherty and
> Takacs, 2004: 573)

They continue:

> Play is how people generate heedful interrelating for innovation in
> the first place, while heedful interrelating is what makes play serious
> enough to be an instrumental style of action for managing large, com-
> plex organizations. (Dougherty and Takacs, 2004: 585)

Consistent with the work of Dodgson, Gann, and Salter (2005), and of Anderson (1994), Dougherty and Takacs (2004) define play in affirmative, even liberating terms: "Play suggests behaviors that are voluntary, pleasant and liberating.
'When we play, we see no limitations'" (Sutton-Smith, 1997, cited in Dougherty
and Takacs, 2004: 575):

> Research reports people perceive 'work' to be constrictive, structured,
> tedious, difficult and boring, while 'play' is seen as liberating, unstruc-
> tured, refreshing and emotional, and suggests that the benefits of play
> can be achieved by framing activities, since relabeling tasks as play in-
> stead of work transformed people's perception, judgments and motiva-
> tions. (Dougherty and Takacs, 2004: 576)

Play, then, represents the absence of limitations, of administrative, cognitive,
and economic limitations, and operates in a sphere where all things are possible,
where all thoughts and ideas have a value, and where everyone can make a contribution. In order to interrelate heedfully in multi-functional teams, an element
of play needs to be in place, suggest Dougherty and Takacs (2004):

> [P]eople interrelate heedfully when they act carefully, critically, wil-
> fully and purposefully with regard to the joint situation, rather than
> habitually. If interrelations are heedful, people are mindful of the 'big
> picture' to which they contribute a part, so their situated activities are
> more likely to also integrate with activities of others in that particular
> situation. (Dougherty and Takacs, 2004: 574)

In the innovation management literature, play and playfulness enable individuals,
or groups of individuals of a diverse background, to open up to new alternative
perspectives. Being cognitively stuck in economic and practical considerations
rarely produces any innovative ideas; however, an open field of interactions and

exchanges embedded in a playful attitude is potentially the best way to achieve creative situations. For instance, in Blau and McKinley's (1979) study of award-winning architect firms (i.e., by definition innovative and creative companies within the industry), it was found that the "[e]xcessive subdivision of tasks and responsibilities reduces the flexibility and openness that are required for highly creative work", i.e., rather than specializing in a few tasks, the co-workers developed general architectural competencies and became highly skilled in what Blau and McKinley (1979) call "social integration", the joint collaborative efforts needed to produce innovative work. In addition, Blau and McKinley (1979: 216–217) also suggest that innovative architecture firms were less concerned with "maximizing efficiency and profit" than with "turning out unique, aesthetically or technically notable projects". "These firms", argue Blau and McKinley (1979: 216–217), "rarely standardize design concepts from project to project, and attempt to continually to evolve new and creative solutions to particular problems." Expressed differently, the innovative architect firms did not serve a uniform market with "routine productions" but emphasized their capacity to create innovative products on the basis of a playful use of all their competencies in-house. The routine-based production of the non-innovative firms did not care to produce creative solutions, instead caring for bottom-line results; however, the innovative firms took pride in producing new architecture. In one of the categories, play is left out, while in the other, it lies at the very heart of operations. Play creates the new.

Summary and conclusion

In the humanities and social science literature, the concept of play is treated as constitutive of more intellectual forms of thinking. Play and playfulness serve a number of functions and roles in undermining taken-for-granted beliefs and assumptions in order to pave the way for critical reflection. For instance, the puns of Groucho Marx, addressed in many settings and from alternative theoretical perspectives, are illustrative examples of how jokes may be used to undermine and call into question the institutionalized modes of using words and expressions. Everyday conversations rest on joint and relatively stable agreements regarding the meaning of words and phrases (syntax and semantics), and the rules of conversation may only be temporarily bracketed without undermining the entire conversation. That is, only very seldom are there possibilities for Groucho-isms to make us aware of how much of our everyday language is taken for granted. Still, this playful undermining of *what is* in order to reveal *what may become* is of great importance for everyday thinking, for providing a glimpse of the existing alternatives. During the medieval carnival, the entire social order is temporarily overturned and is only reestablished after a period of turmoil and disorganization. Similarly, in organizations, say, in innovation projects, the capacity to collectively overcome and de-familiarize predominant social conditions is a most valuable capacity, the principal driver when it comes to enacting new ideas and objectives. Such a de-familiarization is not embedded in calculative practices, and the choice between a series of alternatives, but is, on the contrary, the venturing beyond such stale conditions, a form of joyful capacity to think the new. Therefore, innovation work is not, in essence, calculative and nor does it merely draw on instrumental rationalities; it is, on the contrary, grounded in play and playfulness. As a consequence, there is an endemic and very well researched antagonism in the creative industries between the "creatives" and "the suits" whereby "the suits" are treated as a money-grabbing and finance-minded professional cadre with little understanding of the creativity employed when creating new products or services (Caves, 2000; Hesmondhalgh, David & Baker, 2008; Townley, Beech and McKinlay, 2009; Gotsi, Andriopoulos, Lewis, and Ingram, 2010; Brown et al., 2010). Expressed differently, these two categories of professional workers represent two distinct situational rationalities, two modes of economic agency; "the creatives" draw on the capacity for playfulness, of key importance when conceiving of new ideas, while "the suits" (managers) operate on the basis of a calculative mode of thinking. Using a term from epistemology, these two modes of thinking are "incommensurable" – i.e. they operate along divergent lines of thinking. As a consequence, there are longstanding controversies between "making money" and "being creative" within the creative industries. As contemporary capitalism is becoming increasingly knowledge-based, these controversies may be expected to escalate. However, an understanding of the role of the various situational rationalities at work may mediate such conflicts.

Chapter 3.

Reciprocity

Exordium

In a study of assisted fertilization clinics and the technological and scientific development of reproductive medicine, the interlocutors, a blend of professors, scientists, entrepreneurs, clinic directors, and other professional groups, strongly emphasized that one of the reasons why Sweden and Scandinavia were more broadly at the very forefront of assisted fertilization was that the public healthcare sector has appropriated assisted fertilization as a component of the public heathcare offering. Another important factor in the advancement of know-how and technology was the close relations and collaborations between the public clinics, hosting much of the research work, and the private clinics. "Swedish IVF ... is highly qualified. We collaborate [across the private-public boundary] and maintain a transparency in what we are doing, and there are joint research activities ... In other countries there are more of closed communities [In each clinic]. That doesn't promote any development", argued a gynecology surgeon and CEO of a major assisted fertilization company. One of the directors of a clinic, with long-term experience of the field and one of the pioneers of the industry, recalled how know-how was shared across institutional and corporate boundaries:

> In the late 1970s, early 1980s, we were trading know-how not through published journal papers but verbally. We were travelling around between clinics, looking at their work and trying to change things for the better ... This clinic was at an early stage connected to the University Med School clinic and they were communicating closely, that is, what they did we also did. (Director of Clinic)

One of the reasons for openly communicating research findings was that there was excess demand for assisted fertilization treatment – the industry is estimated to have grown annually by six to seven percent – and the need to establish assisted fertilization as a legitimate and safe therapy. Consequently, the early years of assisted fertilization, from say the late 1970s to the early 1990s, were characterized by collaborative efforts. The director of the clinic described how he had spent a lot of time in the mid-1980s teaching other clinics how to use one of the technologies that he and his colleagues had developed:

> Among other things, we were the first clinic in the world to develop and use this specific technique. I was invited to travel around the globe to demonstrate it … It was a method enhancing egg retrieval. We were the first clinic in the world to use it and we were happy to share our experience. (Director of Clinic)

The history of assisted fertilization, and its gradual institutionalization in the field of healthcare, can be told from many perspectives and along complementary storylines. One such story emphasizes how collaboration and reciprocity played a role in instituting reproductive medicine.

Introduction

> "An object is valuable in relative to a given action system"
>
> Gibson (1977: 79)

In the classic Marxist analysis of capitalism as a historically contingent economic regime, capitalism is characterized not only by the circulation of capital in the processes where capital becomes commodities that are translated into yet more capital, but also in terms of capital no longer merely being the medium for economizing various resources and assets, a means of enabling exchange and reducing what Ronald Coase (1937) would eventually come to call transaction costs; capital is what *precedes* the actual production of goods and commodities. The starting point for economic action is no longer human and social needs but capital *per se*. Such a reification of society in the dry and abstract concept of capital has been a standing criticism from Marxist and conservative scholars alike. In a capitalist regime of economic accumulation and regulation, it is no longer humans who are in charge and who serve as the "measures of all things"; increasingly, it is capital that is becoming the center of relations. In such a situation, it is the free and unregulated market that is the principal *topos* for economic exchange and, in the formal and theoretical enactment of the concept of the market in economic theory, the market is a smoothly-functioning mechanism for setting prices and for almost effortlessly adjusting supply and demand. As the sociologist Neil Fligstein (2001) points out, the difference between economic sociology and economics lies, *inter alia*, in the lack of belief in all markets necessarily being efficient, but that market efficiency is a social accomplishment based on regulatory con-

trol, legislation, and ethical standards enacted by relevant actors, rather than being the starting point for analyzing markets:

> Economic theories start with the premise that social institutions would
> not persist if they were not efficient. In the new institutional econom-
> ics this assumptions is not meant to be tested. Instead, the general
> research tactic is to examine situations where different amounts of
> sources of uncertainty exist and then predict whether or not one will
> observe a certain social relation … Sociological theories are more de-
> scriptive and usually agnostic or skeptical as to the ultimate effect of
> social structures on efficiency. I, too, doubt that all social structures
> are efficient. (Fligstein, 2001: 9)[19]

Guillén *et al.* (2002) speak of "three fallacies" in economic analysis which economic sociologists seek to avoid. The first fallacy is that economists tend to separate the realm of "the social" from the realm of "the economic". In addition, economists, argue Guillén et al. (2002: 7), "generally believe that individuals make
conscious calculations about how to maximize utility, and that the preferences
that determine their utility function (that is, the sources and associated magnitude of their utilities) are exogenous to the models of interest". This view is the
second fallacy of economic analysis according to economic sociologists. Thirdly
and finally, Guillén et al. (2002: 7) identify differences between individual level
behavior and group behavior:

> A final key theoretical difference is that economic sociologists reject
> the idea of the aggregation of individual level behavior as straight-
> forward and unproblematic. Drawing on previous research on social
> classes, social movements, social networks, power dynamics, and cul-

---

19 David Stark (2009) proposes here an alternative to the institutional theory perspective in advocating a theory of entrepreneurship whereby the entrepreneur actively seeks to exploit the unequal distribution of information within a field (see, for instance, Knight, 1921). In Stark's (2009:
184) view, institutional theory has contributed a shift in focus from action as "choice" and "decision" to "scripts and routines as resources for action", i.e., decisions are less a matter of free
choice than habitual action embedded in conventions and shared beliefs. However, Stark (2009:
184) also says, "[t]here is a risk [in institutional theory] that *cognition* becomes reduced to the
'taken-for-granted'. The challenge is to retain the insight that action should not be reduced to
'choice' or 'decision' without reducing cognition to 'unreflective activity'." Stark speaks here of a
"reflexive sociology" recognizing the capacities for reflexivity in the actors studied. Stark (2009)
wants to avoid the Scylla of free choice and rational action endorsed in formalist economic theory and the Charybdis of institutionalized, habitual and unreflected behavior in some quarters of
social theory and, more specifically, mainstream institutional theory by making the entrepreneur
the reflexive actor who creates "[d]isruptions that prevent path-dependent effects of locking in
to early successes" (Stark, 2009: 5).

tural blueprints … economic sociology seeks to improve our understanding of economic behavior at a level of analysis higher than that of the individual or the group. (Guillén, Collins, England, and Meyer, 2002: 7)

Based on these three fallacies, theoretical constructs such as the *homo oeconomicus* ideal type model and rational choice theory are generally treated as *specific* rather than *general* theories of economic behavior: "Empirical evidence has shown that economic action is often not based on a self-interested assessment of incentives, as argued by economists. Instead, it is often based on trust, which is historically developed and culturally specific, although not exclusive to any one culture" (Guillén, Collins, England, and Meyer, 2002: 8).

Bourdieu (2005) is another renowned sociologist who is highly critical of the doctrines of economic theory that poorly respond to actual conditions and empirical evidence but seek to transcend such petty concerns in the construction of a glorious theoretical framework. Bourdieu (2005: 1) happily quotes Bertrand Russell, suggesting that "while economics is about how people make choices, sociology is about how they don't have any choice to make". Here, Bourdieu calls for a critique of economics in terms of dissociating a series of very central choices from the wider social setting wherein the actor is located, namely the surrounding society: "The science called 'economics' is based on an initial act of abstraction that consists of dissociating a particular category of practices, or a particular dimension of all practice, from the social order in which all human practice is immersed" (Bourdieu, 2005: 1). For instance, the concept of *homo oeconomicus*, "economic man" capable of making informed choices between alternatives on the basis of strictly individual cognitive capacities, of key importance to what Bourdieu (2005: 209) calls "economic orthodoxy", is "a kind of anthropological monster" and "the most extreme personification of the scholastic fallacy". *Homo oeconomicus* may serve a role in the economic *Gedankenexperimente* but to take this ideal-typical model as the fundament for economic theory is a violation of the inductive methods applied in the social and economic sciences, based on empirical evidence and inferences made on the basis of data. Here, Bourdieu calls for sociological or social understanding of economic fields such as markets (White, 1981; Swedberg, 2005). For instance, rather than assuming that the preferences, needs, aspirations, and propensities of the individual actor – the abstract model of *homo oeconomicus* – are universal and exogenous, part of some "universal human nature", these are, on the contrary, "*endogenous* and dependent on a history that is the very history of the economic cosmos in which these dispositions are required and rewarded", suggests Bourdieu (2005: 8). Understanding the human agent is thus a matter of first overcoming the "ahistorical vision of economics" and understanding agency not as a wholly autonomous practice, but as something that is, of necessity, bound up with a variety of social conditions and actors within the setting of a particular society.

The concept of reciprocity

"Any sentence that begins 'All societies have…' is either baseless or banal", claims the anthropologist Clifford Geertz (2000: 135). In making a choice between two evils, it is arguably better to pursue a banal argument rather than a baseless one. Thus, to say that all societies have mechanisms of reciprocity may be a banality, but this is simultaneously grounded in extensive research in, for instance, anthropology and social science and thus it is not exactly baseless. Anthropology is inextricably bound up with colonialism and with the exploitation of the economic resources provided by the colonies laid under European governance from the late Renaissance period and onward (Prasad and Prasad, 2003, 2002). In the nineteenth century, the real period of colonial expansion and the growth of international trade, the social sciences were born. During that century, sociology, criminology, demographics, economics, and anthropology were established as disciplines in their own right (Porter, 2009). Theology had to compete with a variety of new scientific disciplines that are competing to explaining human society. Society is no longer, then, an instituted order prescribed by God, but is becoming a man-made arrangement of rules, roles, and norms. Anthropologists were educated and trained in the great colonial powers and sent out to the colonies to study the indigenous people under colonial rule (Prasad and Prasad, 2002). In some cases, much of the work done was armchair thinking based on the reporting of others – for instance, Emile Durkheim's (1995) seminal work on the elementary forms of religious life belongs to this category – but the first generation of anthropologists, such as Bronislaw Malinowski and A.R. Radcliffe-Brown, developed the discipline to include longer periods of fieldwork in foreign and remote territories, naturally accompanied by a minimum of necessities such as furniture and servants that would provide a decent level of comfort while in the field. Anthropologists of the early generations were still in many ways highly receptive to local customs and societies; even though their foundational work was colored by European beliefs and customs, these anthropologists were capable of identifying interesting local customs and traditions, suggesting that indigenous, even so-called "primitive", people shared many mechanisms and structures with the "civilized" Europeans. This enabled a theorizing of cultures as, on the one hand, particular but also anchored in a few recurrent taboos or institutions. One of the most general taboos, and arguably the basis for the principle of reciprocity, is the incest taboo and the principle of exogamy (Lévi-Strauss, 1985: 34–35), that is, the exchange of daughters between families. Based on such generic institutions, anthropologists acquired data and information enabling rather sophisticated theories about the nature of human society. Perhaps the single most important concept developed in anthropology is the concept of *culture*. The German word *Kultur* has been put forth as a compliment to the Latin concept of *civilization* whereby the former denotes a certain quality of differentiated customs, modes of thinking, media for sharing know-how, and so forth, while the latter takes on the meaning of a particular advanced system of law, governance and administration. In contrast to civilization, culture is a term that denotes the norms, beliefs, and worldview of a social formation in its entirety. The concepts of reason and rationality are embedded in such cultures. Lévi-Strauss (1985) suggests that culture needs to be understood as "rules of conduct" that are commonly obeyed:

> Culture is neither neutral nor artificial. It stems from neither genetics nor rational thought, for it is made up of rules of conduct, which were not invented and whose function is generally not understood by the people who obey them. Some of these rules are residues of traditions acquired in the different types of social structure throught which, in the course of a very long history, each human group has passed. Other rules have been consciously accepted or modified for the sake of a specific goal. Yet there is no doubt that, between the instincts inherited from our genotype and the rules inspired by reason, the mass of unconscious rules remains more important and more effective; because reason itself, as Durkheim and Mauss understood, is a product rather than a cause of cultural evolution. (Lévi-Strauss, 1985: 34)

Rules of conduct are thus the elementary structures of social life, a form of basic mechanism regulating social life and leading to the establishment of institutions. Here, Radcliffe-Brown (1958) separates *social structure, social organization,* and *institution.* Social structure "refers to an arrangement of parts or components related to one another in some sort of unity" (Radcliffe-Brown, 1958: 168). In social structure, "structure consists of the arrangement of persons in relation to one another". Social organization, on the contrary, is not the "arrangement of persons" but the "arrangement of activities", suggests Radcliffe-Brown (1958: 169). Finally, an institution is "[a]n established or socially recognized system of norms or patterns of conduct referring to some aspect of social life" (Radcliffe-Brown, 1958: 174). Social structure is the arrangement of individuals and components; social organization is the arrangement of practices; institution is the norms and patterns of conduct. Radcliffe-Brown is not particularly explicit on this point but when we move from social structure to social organization to institution, we move to increasingly differentiated social systems that are mutually constitutive; that is, social organization resides in social structure but also leads to institutional arrangements. In Adorno's (2000: 105) view, institutions are "congealed action", something that has become "autonomously detached from direct social action" yet still serves to coordinate and structure action: "[T]he social action of each individual person … is dependent on these institutions, and can only be explained in terms of them", says Adorno (2000: 105–106). Mary Douglas (1986) goes even further, claiming that institutions are constitutive of thought, and that institutionalized norms and beliefs provide us with a mode of thinking that cognitively structures the world.

For anthropologists and sociologists of the early generations – the term sociology was first coined by August Comte in 1839 (Hobsbawm, 1975: 261) – it was of great importance to examine social practice as embedded in generic social schemata (e.g., Radcliffe-Brown's tripartite model) in order to identify the underlying texture of social behavior. Strongly influenced by the sciences of the nineteenth century, seeking to uncover the "natural laws", the social sciences were nomothetic, concerned about the *nomos*, laws, of society. One of the best-known "sociological laws" was formulated by Émile Durkheim in his *Elementary form of religious life* (first published in 1912) where he distinguished between the *sacred* and the *profane* as two distinct spheres and mechanisms – Durkheim speaks of *genera* – in all societies:

> Whether simple or complex, all known religious beliefs display a common feature: they presuppose a classification of the real or ideal things that men conceive into two distinct classes – by two distinct genera – that are widely designated by two distinct terms, which the words *profane* and *sacred* translate fairly well. The division of the world into two domains, one containing all that is sacred and the other all that is profane – such is the distinctive trait of religious thought. Beliefs, myths, dogmas, and legends are either representations or systems of representations that express the nature of sacred things, the virtues and powers attributed to them, their history, and their relationships with one another as well as with profane things. Sacred things are not simply those personal beings that are called gods or spirits. A rock, a tree, a spring, a pebble, a price of wood, a house, in a word anything, can be sacred. A rite can have sacredness; indeed there is no rite that does not have it to some degree. (Durkheim, 1995: 34–35)

"[T]he sacred and the profane are always and everywhere conceived by the human intellect as separate genera, as two worlds with nothing in common", adds Durkheim (1995: 36). Here, Durkheim is making a universalist claim that all societies are separated into two domains governed by different institutional beliefs and norms.

Marcel Mauss, Durkheim's son–in-law, and one of the greatest French anthropologists, published his study *The Gift* in 1954, suggesting that the principle of reciprocity is central to all societies. The institution of giving gifts between tribes, clans, families, and members of society is a circulation of goods and services that formally does not demand any counter-gift while, in practice, such a counter-gift is institutionalized as something that maintains and upholds social relations and prestige. The incest taboo in virtually all societies, embedded in the biological need to extend the pool of genetic material in order to create viable offspring, leads to the institution of exogamy, in many cases allowing marriages between cousins but not between closer relatives such as siblings.[20] As an extension of this "gift economy" of daughters and sisters, trading between tribes is based on similar principles. Bronislaw Malinowski accounted for the trade between tribes and clans in his *Argonauts of the Pacific*, a specific tradition Malinowski referred to as *kula*. However, there is a difference between a conventional trade relationship, based on economic principles (e.g., the division of labour, calculative practices), and the gift-relationships, which in essence are non-economic (i.e., non-utilitarian) but social in nature. At the same time, implies Mauss (1954), the gift relationship is the basis of continuous economic relations as social relations create confidence and trust between and within a community. In order to sustain a society, goods and services *qua* gifts need to be circulated. In historical and contemporary societies, there is much evidence of gift-relationships, ranging from the (elementary) custom of lighting the cigarette of the person with whom one is smoking to the (differentiated and technoscientific) donation of organs and renewable biological resources such as blood and sperm.

Mauss' study is filled with illustrative cases. Still, the generic principle of reciprocity, of giving and counter-giving, remains of central importance in social settings. While the economics view of economic endeavors enacts a calculative and utilitarian model of agency, in short the *homo oeconomicus* ideal-type, economic sociologists see instituted norms such as gift-giving as important and not as residual factors to consider when examining social relations and economic action. In this latter view, everyday economizing is not a matter of understanding the rationality of socially autonomous individuals but of exploring the totality of norms, beliefs, institutions and traditions wherein such – perhaps seemingly – autonomous calculative practices are located. That is, economic sociologists do not refute

---

20 The institution of marriage has served a key role in most societies in order to create lasting bonds and mutual obligations between families. Padgett and Ansell's (1993) study of the Medici family in Renaissance Florence underlines joint economic interests in business ventures and marriages between families as the two generic principles for the emerging Florentine power, at its height during the first half of the fifteenth century. Thus, the principle of exogamy is not only derived from the need for biological diversity in a population but is also part of the sociogenesis, the development of durable and solid social arrangements.

the very idea that individuals may consciously attempt to maximize their benefit or yield but assume, contrary to many economists, that thoughtful and reflective economic agents take the various norms and beliefs into account when seeking to maximize personal benefits. This means that, at times, there is a need to act against one's own immediate short-term interests in order to benefit from being part of long-term social relations. In what follows, a few examples will be provided of how the principle of reciprocity operates in organizational settings. Rather than focusing more narrowly on the more restricted practice of gift-giving, reciprocity is treated here as a generic term denoting all kinds of mutual affirmation and recognition of the competencies and expertise constituting social and organizational life *per se*. However, the very idea of gift-giving has been of immediate interest in some fields, for instance in industries where lavish Christmas gift to, for instance, suppliers or sub-contractors is a common routine; or in the pharmaceutical industry where the financial support of physiology and medicine congresses and conferences is a standard arrangement, testifying to the intricate relationship between the healthcare system and academic medicine, on the one hand, and the pharmaceutical industry, on the other (Brody, 2007).

The social embeddedness of markets

Empirical studies of the regulation and control of markets support Fligstein's (2001) and Bourdieu's (2005) critique of economic orthodoxy. Podolny (1993) stresses that status and prestige play a key role in markets and, since there is only a loose coupling between the perceived and actual quality of a product, high-status actors get a higher margin on their investments, all things being equal: "The greater one's status, the more profitable it is to produce a good of a given quality" (Podolny, 1993: 841). Podolny (1993) emphasizes that, while economists take differences in product quality as their point of departure, economic sociologists are concerned about understanding the influence of status within the industry prior to the launch of the actual product; for economists, status is the *effect* of superior quality while for economic sociologists, the perceived superior quality of a product is an effect of status hierarchies already in place in the market (see also Podolny, 1994; Rao, 1994). Such "market inconsistencies" are often ignored or overlooked by economists. Studies of rating systems issued by so-called market intermediaries, a key mechanism when it comes to classifying and categorizing both products and market actors, demonstrate that, while such rating systems are promoted in terms of being based on neutral and autonomous professional analyses, a kind of "view from nowhere" if you like, they are in fact mirroring the underlying status and power differences in an industry or institutional field. Market intermediaries serve the market to commensurate assets, that is, to transform "[d]ifferent qualities into a common metric" (Espeland and Stevens, 1998: 314).

Market intermediaries make initially heterogeneous resources comparable and tradable. "Commensuration can be understood as a system for discarding information and organizing what remains into new forms. In abstracting and reducing information, the link between what is represented and the empirical world is obscured and uncertainty is absorbed", explain Espeland and Stevens (1998: 317). Market intermediaries thus serve to reduce transaction costs as they inform choices regarding what asset to buy, hold, or sell. Consequently, their commensuration (e.g., rating of assets) needs to be unbiased and presupposes what Porter (1995a) calls "a trust in numbers" (see also Espeland and Stevens, 2008). Studying the role of classification in financial markets, for instance, in the case of recommendations about whether to buy, hold or sell a financial asset, Fleischer (2009) suggests that such classifications are never devoid of interests:

> Ambiguous classification schemes, blurring boundaries among objects rather than distinguishing them, can allow the organizations that create them to hide behind perceptions of audiences while protecting their own interest. In such cases, the information value of the classification scheme is suspect. (Fleischer, 2009: 556)

"A rating system", says Fleischer (2009: 558), "must strike a careful balance among the interests of the rating organization, the producers of classified products, and the audience of the system." Fleischer (2009: 571) found that the classifications used by market intermediaries contained overlapping categories, "increasing the ambiguity of the classification scheme", and, consequently, market intermediaries influenced the perception of the financial market in "substantial ways": "Although market intermediaries are typically considered impartial facilitators of exchange, they are also self-interested actors who can influence the structure of the information they share in subtle, but substantive ways" (Fleischer, 2009: 573). In a study of the classification of films in the US film industry, Waguespack and Sorenson (2011) found similar market inconsistencies inasmuch as the rating of films was supposed to rest on a professional and disinterested analysis and evaluation of actual scenes that included violence, nudity, and so forth, while in fact prestigious film studies and members of the MPAA [Motion Pictures Association of America, a film producers' association] received more favorable ratings for their films, all other things being equal. "Our analysis of films from 1992 to 2006 confirms that producers' identities matter in the rating of their films: films distributed by MPAA members, and those with highly central producers and directors receive less restrictive ratings", conclude Waguespack and Sorenson (2011: 549). Such rating systems again violate the institutional framework from which they draw their legitimacy and influence markets negatively inasmuch as certain prestigious and high-status actors gain advantages over other actors: "To the extent that high-status individuals and firms have greater freedom to deviate from these restrictions while still receiving desirable classifications, they can pursue profitable strategies not open to lower-status participants, thereby further stratifying outcomes", contend Waguespack and Sorenson (2011: 541). As Fleischer (2009) as well as Waguespack and Sorenson (2011) demonstrate, even with the presence of so-called market intermediaries, established to safeguard "effective markets" through the issuing of neutral and disinterested assessments of products competing in the market, the influence of status and prestige, in turn an effect of power relationships within the industry, is substantial. In other words, rather than making the formalist assumption that markets promote free and unbiased competition, i.e., actors make their decisions and choices strictly on the basis of economic conditions, social factors are always part of any economic transaction already. Thus, formalist theories of economic activity, reducing social action to strictly calculative operations, poorly reflect actual conditions and practices.

Processes of valuation and commensuration: Negotiating worth

In cases where economic interests intermingle with political processes, it is also complicated to separate economic and social interests. A fairly substantial literature in economic sociology and the social sciences more broadly addresses what is called "economic worth" and processes of "valuation" whereby various social resources are inscribed with economic value (Stark, 2009; Aspers, 2009; Garcia-Parpet, 2007; Yakubovich, Granovetter and McGuire, 2005; Podolny and Hill-Popper, 2004; White, 1981; Geertz, 1978). Beckert (2009), criticizing "general equilibrium theory" as the "heart of neoclassical economic theory" and theories of markets, for resting on simplistic assumptions, emphasizes three distinct analytical aspects of markets: (1) the "value problem", (2) the "problem of competition", (3) the "cooperation problem". The first problem occurs when market actors are confronted by uncertainties when assessing the "value of commodities" (Beckert, 2009: 253). Such difficulties are salient in markets with few "objective standards of quality assessment" (Beckert, 2009: 254), for instance the market for contemporary art. In such cases, Beckert (2009: 255–256) proposes, "social belief systems", i.e., institutions, shape the assessment of the commodities and reduce the level of uncertainty in the market. Velthuis' (2003) study of the contemporary art market in Amsterdam and New York shows for instance that gallerists were very concerned not to lower art prices even during a recession because artists valued their work on the basis of its market value and because art collectors become nervous when prices slump. The social belief system thus enacted artists in terms of being closely bound up with the local art market and the prices paid for contemporary art. Regardless of the dominance of such economic reason, Beckert suggests that the institutions constituting and regulating markets are to be "[u]nderstood not from a contractarian perspective as the efficient result of an agreement of socially unbound individuals, but rather as situated within a specific political, social and cultural context that constitutes the actors' goals, strategies, and cognitive orientations" (Beckert, 2009: 250–251). As a consequence, "institutions are ... not to be understood as efficient responses to information problems", says Beckert (2009: 250). When there are difficulties assessing economic value, institutions play a central role in determining costs and benefits, i.e., in making economic valuations.

Fourcade's (2011) study of the commensuration and valuation of nature in the event of oil spills off the coast of Brittany in France and in Prince William's Sound in Alaska show that there are no clearly-defined and rational models for pricing and valuing such "priceless resources". As a consequence, the process of the economic valuation of what Fourcade calls "peculiar goods" becomes a political one:

> [I]f money is one, monetary commensuration (or economic valuation) techniques are numerous and varied. The production, selection, and application of these techniques is thus extraordinarily contingent and deeply political, raising questions about 'scientific trials of strength' and the processes of 'translation' and 'allies of enrollment' (Latour, 1987) that stand behind them. (Fourcade, 2011: 1725)

"Accounting techniques are social constructs too: they emerge and gain authority in particular systems of expertise, social relations, and cultural narratives prevalent in these contexts", says Fourcade (2011: 1728). Lakoff and Klinenberg (2010), studying the use of rule-based calculation to allocate economic resources from the Department of Homeland Security in the US, also emphasize that there are no scientific and impersonal methods for allocating financial resources prior to the political process: "[R]ule-based calculation as the basis for bureaucratic decisions is an ideal rather than a description of typical practice. Only under certain circumstances are bureaucratic agencies actually required to publicly demonstrate that they are following calculable rules" (Lakoff and Klinenberg, 2010: 506). The Department of Homeland Security was criticized by politicians for making political considerations rather than reviewing actual risks: Wyoming received as much as $61 per capita while California only received $14, and sparsely-populated states (dominated by Republican voters, representatives of disfavored states were quick to remark) such as Montana and North Dakota consistently received more money per capita than, for instance, New York and New Jersey. However, the practices for calculating risks had never been agreed upon; consequently, there were "no impersonal means to adjudicate such a dispute" (Lakoff and Klinenberg, 2010: 512):

> Without agreed-upon standards of measurement, it was impossible to
> compare the relative level of threat and vulnerability of various cities
> and regions, and so there was no definitive way to determine whether
> homeland security funds were being allocated objectively, that is, in-
> dependent of 'political' considerations. (Lakoff and Klinenberg, 2010:
> 512)

In other words, "transparent and shared standards of measurement must be in place to make objective, 'de-politicized' calculation possible", suggest Lakoff and Klinenberg (2010: 522). Allocating shared resources, then, is not based on some "extra-social" or universal economic reason, but needs to be negotiated and agreed upon. Rule-based calculation as a bureaucratic routine is embedded in political, i.e., social and cultural, processes. Lakoff and Klinenberg (2010: 523) conclude: "From this case, we can see that in contemporary struggles over security resources, the distinction between 'rationality' and 'politics' does not exist *a priori*, but rather is defined in the concrete political and technical struggle over the creation of a decision tool." Fourcade makes a similar argument, pointing to the impossibilities of commensurating, for example, natural resources separated from the social and cultural recognition of such resources:

> Economic valuation is so revealing precisely because it is so much
> more than a process of monetary commensuration: it is, much more
> powerfully, a process of 'definition' or social construction in a substan-
> tive sense (Smith, 2007) which incorporates all kinds of assumptions
> about social order and socially constructed imaginaries about worth.
> Economic valuation, in other words, does not stand outside of society:
> it incorporates in its very making evaluative frames and judgments that
> can all be traced back to specific politico-institutional configurations
> and conflicts. (Fourcade, 2011: 1769)

For Fourcade, the practices of valuating nature reveal that nature is "never just nature" – which is a cultural construct with its roots in nineteenth century modernism – but an "an assemblage of relations involving humans and nonhumans" including formal and bureaucratic procedures such as "state policies, legal rules, political commitments, economic technologies, and ecological theories" and also constituted in the everyday uses of nature, i.e., "in the strolls taken along the coastline, the shellfish collected for dinner, the ways of life of fishermen and the sand walked on by visitors, in the claims of scientists and the policies of public officials, and in the ideas and emotions that landscapes we may never have seen evoke in the presumed 'public' – us" (Fourcade, 2011: 1770).

Fourcade (2011) as well as Lakoff and Klinenberg (2010) show that commensuration and economic valuation, as two economic procedures embedded in calculative and economic reason, can never be understood in isolation from the social and cultural setting that legitimizes and justifies how economic worth is calculated. Again, the very idea of a transcendental economic order is an ideological construct and a fabrication. Economic reason, no matter how sophisticated formalist models and theoretical frameworks are constructed, is always folded into social and cultural concerns.

RECIPROCITY IN ORGANIZATIONS

Relationality in organizations

The term relationality has been introduced into organization theory to denote the importance of close-knit social relations in organizations supporting and reproducing identities and forms of knowledge sharing (Cunliffe, 2008; Cooper, 2005). Lee and Brown (2002: 259) speak about the "decentering of the subject", the enactment of the subject as an intersection of forces and possibilities rather than an autonomous and enterprising individual. This is a central theme in continental philosophy during the post-WWII period in the writings of Jacques Lacan, Michel Foucault, Gilles Deleuze, and Jacques Derrida, the foremost representatives of poststructuralist thinking, a body of texts of key importance to what Orlikowski (2010a) calls a *relational ontology*. Lee and Brown (2002) thus suggest that the subject is never wholly separated from external influences and regimes of power and ideology but is instead the most salient effect of such regimes. The subject is thus constituted in a field of forces and is the outcome of a number of relations between various discursive formations. Here, Clifford (1988: 10) is speaking critically about "transcendental regimes of authenticity", claiming that they need to be complemented by, preferably substituted for, identities that are "mixed, relational, and inventive". That is, the subject and his/her identity do not reside in transcendental conditions but rather emerge in the field of the intertextuality of everyday speech and practice. In the study of organization, as constituted by storytelling and narrative, the subject and his/her identity are bound up with the stories being told and, consequently, the subject becomes an internally-unstable and continually-changing entity open to external influences. Cunliffe, Luhman, and Boje (2004: 276), using the term *meaning*, take such a perspective:

> Meaning unfolds in narrative performance, in time and context, as storytellers and listeners discuss their experiences, interweave their own narratives: a polyphony of competing narrative voices and stories told by many voices within different historical, cultural, and relational contexts. (Cunliffe, Luhman, and Boje, 2004: 276)

Everyday life is characterized by polyphony and intertextuality and the subject's identity and meaning are constructed on the basis of such a multiplicity of voices. In Kondo's (1990) seminal study of the constitution of the subject in a Japanese context, such a polyphony of voices constantly participated and intervened in the subject-formation process: "My neighbors, friends, co-workers, and acquaintances never allowed me to forget that contextually constructed, relationally defined selves are particularly resonant in Japan. I was never allowed to be an autonomous, freely operational 'subject'", says Kondo (1990: 26). In the Western, for example Cartesian, tradition of thinking, the self is enacted as the starting point for thinking. "*Ergo cogito sum*", claimed Descartes, "I think therefore I am." But Descartes' systematic doubt would only let him infer that there are thoughts; it does not imply, as has frequently been pointed out, that the existence of thought necessarily warrants the existence of an "I", a thinking, rational subject. In Kondo's (1990) and others' views, the subject is instead a knot in a field of intertextuality constituted by the narratives being told about the subject. In this view, the subject is the outcome of joint agreement regarding the qualities and preferences of the subject; increasingly, subject-positions are not under the full control of the individual, but the joint construction of agency.

Such a relational view of the subject is less attuned to contemporary late-modern ideologies that emphasize enterprising and entrepreneurial capacities, instead offering a more theoretically-subtle image of how, for instance, professional work, or what has been called knowledge-intensive work, is constituted within relational frameworks wherein competence, creativity, or any other favored quality do not exist *per se* but are embedded in the recognition and affirmation of peers instead. A brilliant chess player cannot be regarded as brilliant in isolation but needs to be addressed in such terms by other chess players, some of whom are at the same level of expertise and some of whom are less advanced. The same goes for scientists, athletes, architects, artists, musicians, and so forth. The principal source of recognition for any professional group is recognition by peers, or at least other actors with the competence to pass judgment on the work accomplished. For instance, studies of string quartets, for example, show that it is not the audience but *other* string quartet musicians who are in the position to affirm the accomplishments through the joint efforts of a particular string quartet (Murningham and Conlon, 1991). The audience may have some very informed listeners, expert listeners (see, for instance, Hennion, 2002), but generally it is only other musicians who can fully understand and recognize musical performances. Similarly, scientists may care about what students think or how the wider public receives their work, but what really counts, at the end of the day, is the recognition of peers, fellow scientists. Robert Merton cites the stress researcher Hans Selye in order to underline the anxious concern whether or not to receive peer recognition:

> Many of the really talented scientists are not at all money-minded; nor
> do they condone greed for wealth either in themselves or in others. On
> the other hand, all the scientists I know sufficiently well to judge (and I
> include myself in this group) are extremely anxious to have their work
> recognized and approved by others. Is it not below the dignity of an ob-
> jective scientific mind to permit such a distortion of his true motives?
> Besides, what is there to be ashamed of? (Hans Selye, cited in Merton,
> 1973: 400)

In some industries and sectors, there are end-users who play a key role in recognizing performance. Elsbach (2009) studies the community of toy car designers, producing collectors' items for sale at toy car fairs and similar events; in this
community, the expert observers of the end-users, the buyers and toy car collectors, serve to constitute a "relational identity" of the car designers. That is, the
toy car designers are not in the position to establish viable identities on their
own, but need to receive the approval and recognition of the end-users in order
to think of themselves as toy car designers. Elsbach's study thus shows that various professional communities cannot subsist, contrary to the ideology of professionalism in terms of having an autonomous and self-contained position, without the role of both other professions and, in some cases, end-users. Relationality
plays a key role in situations and fields where there are few objective standards
of performance that undermine the autonomous self-monitoring and self-control
of activities.

In the study of emotional labor, too, the concept of relationality has been emphasized. Although feelings like anxiety are commonly defined as feelings with no
immediate direction or cause, they emerge, in many cases, within specific situations and under determinate conditions. For instance, Theodosius (2006) claims
that emotions are inherently socially-embedded responses to perceived conditions:

> Our emotions and feelings do not just arise in isolation within our-
> selves, they are directly elicited in response to the relationships we
> have with others. Between people, emotion states collide, bouncing
> and feeding off one another, creating a further emotional state born out
> of that interaction. Thus, in the very simple example of the nurse giving
> sympathy she also receives gratitude in return and further responds to
> this – a conscious and socially informed encounter. There are also a
> whole plethora of feelings and emotions unconsciously involved in any
> given interaction. (Theodosius, 2006: 900)

In this view, feelings are never fully "my own" but emerge, on the contrary, in the intersection between the subject and the external world. Thus, in addition, emotions are not universal but acquired within a specific historical and cultural context; what may be deeply humiliating in one culture – say the misspelling of a name or getting a title mixed up – may be just a source of mild amusement in another. In Theodosius' (2006) view, emotions are based on social relationship and need to be examined within their social setting, and not just as psychological responses to perceived conditions.

In summary, the concept of relationality underlines the fact that subject-positions, identities, and more mundane terms such as social status and prestige are all embedded in social relations that presuppose certain forms of reciprocity. Hard work put into one specific field of expertise translates at times, but not always, into social prestige, even admiration; in many cases, this admiration demonstrated by peers may serve as a stronger incentive than what has been called *extrinsic rewards*, e.g. pay and bonuses. The more complicated it is for outsiders to value and pass judgment on the performance of the individual agent, the more that individual will rely on peer-recognition. A skilled hand surgeon of course appreciates the gratitude of patients benefitting from her competence, but what truly counts is the recognition shown by other hand surgeons. Expert and professional work, or any field of expertise for that matter, is inextricably bound up with relational exchanges. The principle of reciprocity lies at the very heart of all forms of expertise, high-brow as well as low-brow.

Labor-management relations and the bureaucratic destruction of reciprocity

Reciprocity remains a key principle in all social life, so too in the domain of organizations. Alvin Gouldner's (1954a) study of the bureaucratization of an American gypsum plant, *Patterns of Industrial Bureaucracy*, one of "the pillars in organizational sociology" (Hallett and Ventresca, 2006: 912), is a study that nicely illustrates the principle of reciprocity. Alvin Gouldner was trained by Robert Merton at Columbia University in New York City in the years following WWII and is a part of the core of the first great generation of American organization theorists that includes, for example, Peter Blau (Scott, 2004). For some, Gouldner remains a key actor in organization theory (Nord, 1992), while for others, he remains a surprisingly marginal figure, partly because of his difficulties in forging lasting relations with younger scholars who would advance his work (Pedraza, 2002). Still, *Patterns of Industrial Bureaucracy* remains a seminal contribution to our understanding of organizations and management in terms of pointing out how procedures of bureaucratization do not establish what Erwin Goffman (1961) called "total institutions" or Weberian "iron cages" of routines and rules, but rather institute relatively flexible and malleable social structures. In the Midwest Gypsum plant studied by Gouldner, the old manager, called Old Doug, had established what Gouldner calls the "principle of leniency", i.e. not being too heavy-handed in his monitoring of his workers' activities and how they followed workplace rules, and always giving workers who failed in any way to follow the rules "a second chance". When a newly-appointed manager named Vincent Peele, a replacement for Old Doug, started to implement a new regime of rules and regulations, the plant workers responded quite negatively and the whole conflict ended in a wild-cat strike that Gouldner (1954b) accounts for in a research monograph accompanying *Patterns of Industrial Bureaucracy*. While Old Doug had established a specific regime of reciprocity whereby the principle of managerial leniency was applied in exchange for the loyalty of the co-workers, Vincent Peele effectively – arguably on purpose – destroyed these reciprocal relations by implementing a series of bureaucratic rules; a managerial regime based on social relations was displaced by a bureaucratic regime of rules. An intricate system of gifts and counter-gifts was overturned by a rule-governed system of control. Gouldner (1954a) demonstrates how informal gift-relationships are substituted for formal bureaucratic rules and regulations some of which were coercive, some symbolic (leading to what Gouldner calls mock-bureaucratic rules), and some followed without any need for managerial oversight. Bureaucratization lies precisely in the overturning of seemingly arbitrary social relations with more solid bureaucratic rules; in the modernization of society – its "de-traditionalization", one might say in the Weberian tradition of thinking – social relations are not maintained by custom or honor (determined by the capacity to reciprocate gifts), but by contracts and self-discipline. Norbert Elias suggests that the modernization of society is based on the shift in focus from external force, *Fremdzwang*, to self-control, *Selbstzwang* (Bijsterveld, 2008: 250), the diminishing role of gift exchange at the expense of anonymous contracting.

Gouldner's (1954a) study demonstrates that the de-traditionalization of labor-management relations, and the establishment of bureaucratic procedures and rules, do not constitute an effortless accomplishment, rather they overturn or re-formulate social relations on the shop-floor, and in the organization more broadly (i.e., the social relations explored by classic industry sociology research. See, for example, Jaques, 1951; Roy, 1952; Dalton, 1959). Bureaucracy not only reformulates social relations, it also imposes the linear time perspective of instrumental rationality, the temporality of the modern period.

Gift economies in knowledge-intensive work

Alvin Gouldner (1960: 177) argues that functionalist and utilitarian theories are only partially capable of explaining how social interactions are introduced and maintained. In professional and knowledge-intensive domains, in particular, where there is a need to have a speaking-partner or collaborator in order to advance a field of expertise, the shortsightedness of functionalist and utilitarist theories fails to explain many different forms of social interactions which appear to extend outside of the narrower domain of instrumental rationality. Gouldner (1960) thus calls for a change in perspective and an increased emphasis on reciprocity and gift economies. Cheal (1988: 19, original emphasis omitted) defines a gift economy as "a system of redundant transactions within a moral economy, which makes possible the extended reproduction of social relations". Gift economies, the institutionalization of reciprocity as the predominant social norm regulating social exchanges, are observable in many different settings and in domains of both historical and contemporary social organization. Currah (2007) provides a few examples of gift economies:

> [G]ift economies continue to appear in many forms, serving myriad needs and interests in communities of different sizes and geographical reach – for example, Alcoholics Anonymous (and its progeny the Twelve-step programme), blood donation systems, charitable organizations, scientific research or community-led initiatives such as restoration projects. These activities are generally able to operate in a more and efficient and desirable fashion without the alienating and calculating power of the market. (Currah, 2007: 475)

Even though the literature on gift exchange emphasizes that such instituted orders exist in all kinds of societies, Gouldner (1960) stresses that there are also other instituted norms compensating for the lack of gift economies; e.g. when gifts are treated as bribes or are in other ways illegitimate:

> Although reciprocal relations stabilize patterns, it need not follow that a lack of reciprocity is socially impossible or invariably disruptive of the patterns involved. Relations with little or no reciprocity may, for example, occur when power disparities allow one party to coerce the other. There may also be special mechanisms which compensate for or control the tensions which arise in the event of a breakdown in reciprocity. Among such compensatory mechanisms there may be culturally shared prescriptions of one-sided or unconditional generosity, such as the Christian notion of 'turning the other cheek' or 'walking a second mile', the feudal notion of *noblesse oblige* or the Roman notion of 'clemency'. (Goulder, 1960: 164)

In this view, gift exchange is one mechanism among others that upholds and maintains social exchanges (see, for instance, Blau, 1964).

Konstantinou and Fincham (2011) use a Maussian perspective on knowledge sharing, conceiving of the sharing of valuable knowledge as an obligation in a community of 58 knowledge workers in the Greek subsidiaries of five multinationals. When receiving a valuable piece of information, or know-how, from a colleague, this gift was "honoured" through its "application and competent use" (Konstantinou and Fincham, 2011: 839). Such competent use was regarded as a "prime obligation" among the engineers. Working under a constant inflow of information and knowledge, the engineers were also concerned about when to share information and what information to share in order to optimize their work performance; withholding and blocking information was thus also a kind of "gift".[21] In Konstantinou and Fincham's (2011: 839) view, the knowledge gift relationship extended social relations between giver and recipient and developed participation in organizational networks. The study reveals that gift-giving is the sharing of know-how and information but also, somewhat paradoxically, the blocking or filtering of information – the double negation of the "non-gift" of "non-valuable information", one may say – that would disrupt the workflow.

In some cases, social relations derived from ongoing social interactions become an asset that is guarded by actors since such relations are the outcome of social, economic, and emotional investment on the part of the actor and may accrue substantial economic value when handled properly. Darr (2003) studied gift exchange in the electronic markets industry where salespeople maintained their individual stock of clients with whom they exchanged gifts and countergifts. Darr (2003) explains:

> As in the case of lawyers … outside salespeople referred to their relationships with buyers as a tangible asset, which they privately owned … Their personal relations with specific buyers and engineers were a major source of occupational power, and served them as an insurance policy against layoffs, since these relationships were important to their companies and costly to establish. Personal gifts helped in cementing these personal ties. (Darr, 2003: 48)

---

21 Barley, Meyerson, and Grodal (2011) demonstrate that the inflow of e-mail and the fear of both lagging behind in "regular work" and failing to detect valuable information in incoming mail were major concerns for professional workers. Reducing the amount of e-mail is, thus, in itself a valuable contribution, testifying to the "gift" of not sharing what appears to be information of minor importance.

The salespeople commonly handed out samples of new components to their buyers in exchange for information about new design projects. Without these close-knit social relations, the salespeople would not be able to lower transaction costs in the electronics components trade and they would not get access to ongoing discussions within the industry. In the process of gifting and counter-gifting, win-win situations emerge for the two parties. Darr (2003) thus concludes that there is nothing archaic about gift exchange, a procedure reserved for primitive or less economically differentiated societies (Carrier, 1991); on the contrary, gifting is a widespread practice serving many purposes: "[G]ifting occurs closer to the heart of advanced industrial capitalist markets, and is inextricably intertwined with the daily exchange of standard commodities. Commodity exchange in the industrial market for passive components was based on obligation networks and norms of reciprocity created and sustained by way of gifting and countergifting", concludes Darr (2003: 49).

In some fields, the gift economy is less pure and demonstrates a certain hybridity as gifts are at times mixed up with commodities such as what Bird-David and Darr (2009) call "mass-gifts", gifts handed out in retailing settings where the purchase of certain commodities is accompanied by a "free gift". As Derrida (1992), for instance, has discussed, there are no free gifts in a long-term perspective; however, in the short-term perspective, the consumers studied by Bird-David and Darr (2009) felt the obligation to accept the mass-gift:

> Most shoppers collect what they are offered, regardless of what the mass-gift is and whether or not they need it. Some later even take the mass-gift, only to later throw it away, rather than to refuse to accept it. In Mauss's terms, one could argue, most shoppers feel obliged to go and receive the mass-gift which they are offered. (Bird-David and Darr, 2009: 316)

Contrary to the gift economies on the Internet, where the actors had established a relatively predictable instituted reciprocal order whereby programming skills laid down in lines of written code were compensated for by the recognition of peers and prestige, the mass-gift was hybridized and took on what Serres (1982) called a "parasite function", being neither a gift nor a commodity but a hybrid of the two, lying betwixt and between the instituted categories of the regular social order of the market. This liminal position of the mass-gift made customers nervous insofar as they apparently accepted gifts they did not want or need. Bird-David and Darr (2009: 320) still argue that gifts and commodities and hybrids of the two are "intertwined at the heart of mass consumption market" and that the "mass-gift economy" is part of a "sophisticated social fabric of mass markets" that has been explored relatively little in the literature. For instance, in some cases, when customers did not want their mass-gift, they tried to negotiate a price reduction instead, apparently undermining the very idea of the mass-gift as something supplementary and additional to the regular purchase, while simultaneously confusing the difference between the price paid for a commodity by the customer (i.e., its exchange value), the retailer's cost of procuring the mass-gift, and the symbolic value of the mass-gift *per se*. In all cases, the retailer rejected any such negotiations. Interestingly, Bird-David and Darr (2009: 321) end their analysis by suggesting that the use of mass-gift needs to be understood within "the economy of excess" theorized by Georges Bataille (to be discussed in the next chapter), as a form of systematic and strategic squandering of economic resources in order to accomplish certain social effects (Cheal, 1988).

Feminist organization theorists have also been interested in gift exchanges as a mechanism that regulates organizational activities. Tyler and Taylor (1998) studied the work of flight stewardesses and noticed that there were certain relatively rigorous standards regarding how female workers should look, dress, and interact with clients in their work. Such standards, enacting a relatively conservative regime for female workers, argue Tyler and Taylor (1998), institute an ornamental and aesthetic role for the female co-workers of the organization. The airline industry has historically been associated with elites and privileged social strata; consequently, being served by physically attractive female – and male for that matter – stewardesses and stewards is part of the allegedly pleasant experience of air travel, giving a sense of exclusivity. Thus, being able to hire and mobilize beautiful women, suggest Tyler and Taylor (1998), "adds value" to the travel experience. However, the labor demanded of the female stewardesses in terms of, for example, maintaining a slim body and a neat physical appearance any time she is on duty is not insignificant (Tyler and Abbott, 1998; Hochschild, 1983). Tyler and Taylor (1998) say that female stewardesses offer the "gift of beauty" in exchange for being part of an industry that is still – arguably an image somewhat in decline – associated with adventure, glamour, and exoticism. Thus, stewardesses are not only service workers performing a series of activities, from safety instructions to sales work, they also serve the role of acting as the physical embodiment of the airline brand. Like few other industries, the airlines are highly gendered in terms of "hyper-feminine" women and "hyper-masculine" men who perform separated domains of work (Borgerson and Rehn, 2004: 467) – service work (mostly performed by women) and piloting (mostly performed by men). Tyler and Taylor (1998) underline the fact that, in many occupations (e.g., retailing), women are expected to be physically attractive, servile, and (preferably) young, offering what Warhurst and Nickson (2009) refer to as "sexualised labour". The perceived physical attractiveness of individual women and a few men thus becomes part of the gift economy. Williams and Connell (2010) take a similar view – albeit with class being substituted for gender – when examining what they call *aesthetic labor* in upscale retailing, hiring middle-class workers who "look good and sound right" in order to attract middle-class consumers (see also Warhurst and Nickson [2007] on retailing and hospitality work and Wright [2005] on bookstore sales staff). Williams and Connell (2010: 350) define aesthetic labor as "the worker's deportment, style, accent, voice, and attractiveness". "Aesthetic labour", argue Warhurst and Nickson (2007: 107), "is the employment of workers with desired corporeal dispositions. With this labour, employers intentionally use the embodied attributes and capacities of employees as a source of competitive advantage." As retailing is poorly paid in the US (only the minimum wage of around seven dollars per hour is a common wage), these middle-class workers have to rely on others to be able to support themselves and their families. At the same time, Williams and Connell (2010) found that many retail workers appreciated their job and identified with the brand they were selling, not even seeing the job as being worth more money. In some cases, there was cynicism regarding the pay and the quality of the products sold, and Williams and Connell (2010) speak of commodity fetishism as a key explanatory category when understanding retail and how it relies on recruiting overqualified middle-class workers.

In summary, gift economies exist in many different fields and emerge in many different forms as well as in the Internet setting, using digital media; what may be initially regarded as an archaic form of economic exchange is reappearing in new forms. The reciprocal norm of the gift/counter-gift is not, then, simply an indispensable element of economies based on scarcity but also of the domain of information abundance on the Internet. What is being traded here. then, is not a foodstuff or other resources necessary for maintaining the community, but credibility and prestige, helping to establish hierarchies and more solid social arrangements.

Gift relationships in reproductive medicine

Richard Titmuss (1970) uses the concept of the *gift relationship* is in his seminal work on blood donation in the UK and elsewhere. Titmuss is sceptical regarding the commercialization of the blood supply in the healthcare sector and compares the system in the UK, based on voluntary donations, with the one in the US, based on commercial procedures and interests. In Titmuss' view, exploring the effectiveness of the two systems along four parameters – economic efficiency, administrative efficiency, price, and "purity, potency and safety" – the donation system, "the gift economy", is the most effective in all four parameters (Titmuss, 1970: 205). Commercializing the "blood market" means unwanted donors (e.g., drug addicts in need of money) entering the market and, consequently, costs will increase and the quality of the blood will be at stake. Titmuss (1970) concludes that the most efficient way to supply the blood needed in the healthcare sector is to nourish a gift economy where members of a particular society jointly give and receive, when they need it, blood from the healthcare system. Titmuss' (1970) thesis is radical inasmuch as he rejects the effectiveness of commercializing a resource which apparently has great economic value for the end-user and for society in general. Needless to say, the neoliberal doctrines that have been fashionable since the late 1970s have worked against Titmuss' communal view of the supply of blood; today, the "gift economies" in healthcare, e.g. the field of reproductive medicine, are based on a blend of commercial interests and gift-giving. For instance, in the field of oöcyte donation, human egg donation, of great importance in reproductive medicine, gift economies and commercial interests are folded into one another in order to balance two institutional logics.

Mitchell and Waldby (2001) make a distinction between intensive clinical labor (e.g., oöcyte donation) and extensive clinical labor (e.g., blood or sperm donation). While blood donation requires a relatively limited amount of work on the part of the donor – the whole procedure normally takes something like fifteen minutes – the donation of eggs requires hormone treatment from the donor, leading to many not entirely pleasant side-effects. As a consequence, it would be complicated to find egg donors unless there is some economic compensation for the clinical labor involved. Still, egg donors are not, studies suggest, primarily donating because of any commercial interest, but because they wish to help childless couples and single women to get a child of their own. As a consequence, the egg agencies providing the *in vitro* fertilization services are trying to strike a balance between treating egg donation as a "job" and giving "the gift of life". Almeling's (2007) study of egg agencies and sperm banks in the US points to the delicacy of keeping these two institutional logics apart: "[E]gg agencies structure the exchange not only as a legalistic economic transaction, but also the beginning of a caring gift cycle, which the staff foster by expressing appreciation to the donors, both on behalf of the agency and the agency's clients" (Almeling, 2007: 333). She continues: "Agency staff simultaneously tell potential donors to think of donation 'like a job', while also embedding the women's responsibility in the 'amazing' task of helping others" (Almeling, 2007: 334). On the one hand, the egg agencies are companies run on regular economic and financial principles, while on the other, they deal with "life" and thus make money on the trading of human oöcytes while simultaneously having to, for institutional reasons, uphold the idea of donation as separated from such sordid commercial interests that by and large are incompatible with life *per se*. Healy (2004), studying organ donations in the US as an act of altruism, rightly remarks that many theories of altruism emphasize social-psychological explanations but that altruism is also "[s]tructured, promoted, and made logistically possible by organizations and institutions with a strong interest in producing it"; i.e., "altruism … is highly institutionalized" (Healy, 2004: 387). In a way, altruism is an outcome of organizing.

As the economic sociologist Vivianne Zelizer (2005) has pointed out, the intersection between money and intimacy "[g]enerates conflict, confusion, and corruption" and thus people passionately debate "[t]he propriety of compensated egg donations, sale of blood and human organs, purchase of child care or elder care, and wages for housewives" (Zelizer, 2005: 27). For instance, one of the reasons why prostitution (besides a number of reasons pertaining to gender and class issues) is prohibited or surrounded by moralistic concerns is that sexual relationships are inscribed within the institution of marriage and any attempt at creating a market for sexual services means that a series of institutional norms need to be either ignored or violated. As soon as the intimacy of sexual relations becomes subject to economic exchange, two opposing institutional logics and human demands are blended in ways that are problematic for social actors to reconcile. In Sweden, at the forefront of more equal gender relations, the *selling* of sexual services is legal while the *buying* of such services is prohibited by law; legislation that seeks to signal that prostitution is socially undesirable while avoiding to stigmatize women and men who prostitute themselves. To some liberal feminists (e.g., Chapkis, 1997), the prohibition of prostitution is little more than a denial of the agency of women, making the judgment that they prefer to engage in sex work rather than accept other work opportunities they are given because this domain of work pays better and offers other opportunities that they appreciate. For more conservative and left-wing feminists, such liberal accounts of prostitution overrate the agency and enterprising capacities of the women and men in the sex industry, who easily end up in situations where they are exploited.[22] To such criticism, liberal feminists commonly respond that the legislation outlawing pros-

---

22 Another case where the issue of agency and structure has been debated in feminist circles is women's use of cosmetic surgery (see, for instance, Heyes, 2009; Jones, 2008; Pitts-Taylor, 2007; Blum, 2005; Brooks, 2004). Some researchers, like Kathy Davis (2009), argue that their research reveals that women undergoing cosmetic surgery claim that "they did it for themselves" and refuse to regard themselves as having been duped by the cosmetic surgery industry and the beauty industry more generally. On the other hand, Susan Bordo (2009) suggests that it is complicated to just take such claims of agency at face value. Instead, such claims of being and enterprising agent being concerned about one's look is precisely the discourse being promoted by the cosmetic surgery industry. "[C]osmetic surgery is more than an individual choice; it is a burgeoning industry and an increasingly normative cultural practice", says Bordo (2009: 24). While Davis (2009) stresses agency in a liberal tradition of thinking, Bordo (2009) maintains that agency is always constituted already within a specific discourse and field of social practice. Bordo's (2009) Althusserian view of agency (Althusser, 1984), as something entirely different than the liberal idea of agency, underlines the difficulties of assuming self-conscious and enterprising subjects. On the other hand, reducing agency to a set of predefined scripts provided by industry and the authorities is not a meaningful analytical construct. As Davis (2009) says, "Agency is invariably linked to social structures and yet never entirely reducible to them, it is always multilayered, involving complicated mix or intentionality, practical knowledge, and unconscious motives."

titution does, in fact, regardless of its benevolent ambitions, promote such exploitation in terms of obscuring the economic activities pertaining to sex work and giving the authorities little mandate to oversee the industry. If prostitution and other forms of sex work were treated like any other small business (e.g., domestic services or gardening), there would be more opportunities for women and men to take advantage of their interest, liberals claim. Regardless of the position taken in this highly-emotional and politically-charged topic, the very idea of prostitution as a commercial service is illustrative of Zelizer's (2005) general thesis regarding the incompatibility of the concepts of commerce and intimacy. The domestic sphere is governed on the basis of gift and mutual recognition (i.e., love and affection); the public sphere on the basis of contracts and monetary compensation and exchanges.

In Almeling's (2007: 334) study, egg agency staff were highly critical and condescending of women who sought to "make [a] career of selling eggs", partly because such ambitions violate what Almeling calls "the altruistic framing of donation". At the same time, egg agency staff encouraged donors to think of the work as a "job", thus sending mixed signals inasmuch as egg donation is both based on brute commercial interests and altruistic gift-giving. Almeling (2007: 334) also notices that many egg donors were given gifts such as flowers, jewelry, or an additional financial compensation, in order to further extend and uphold the "constructed vision of egg donation as reciprocal gift-giving, in which egg donors help recipients and recipients help donors". In contrast to egg donors, sperm donors, engaging in extensive clinical labor, were not commonly treated as individuals giving "the gift of life"; their sperm had a relatively low status as a commodity in the field of reproductive medicine. In addition, while egg donors were expected to be either attractive and successful (i.e., college students) or to have proved their capacities in the field of reproduction (having one child or, preferably, a few children of their own), male donors were just expected to be healthy and to have an adequate "sperm count" – a good concentration and motility of their sperm. In Almeling's (2007) account of egg agencies, the gift is never fully a gift because it is given to a relatively anonymous couple or to single women whom the donor has only met on a few occasions. At the same time, the relationship between donor, recipient, and egg agency jointly creates a sense of altruistic gift-giving that helps all three parties to reconcile the institutional logics of economic interest and "the gift of life". Thus, eggs are never sold but given, while still involving financial compensation.

Another field of reproductive medicine, surrounded by more controversy and debate than assisted fertilization, is gestational surrogacy, also resting on a gift economy. Wilkinson (2006: 116) lists a few of the common objections to surrogacy, including the commodification of children or (more generally) reproduction, and the exploitation of women, especially those who are poor and vulnerable in some way. Here, Wilkinson (2006: 136) cites feminist theorist Andrea Dworking, treating surrogacy in analogy with prostitution:

> Motherhood is becoming a new branch of female prostitution with the help of the scientists who want access to the womb for experimentation and power ... Women can sell reproductive capacities the same way old-time prostitutes sold sexual ones but without the stigma of whoring because there is no penile intrusion. It is the womb, not the vagina, that is being bought. (Andrea Dworkin, 1983, cited in Wilkinson, 2006: 136)

In Dworkin's view, surrogacy is yet another means of exploiting women, with the relative lack of global legislative harmonization testifying to the controversies surrounding the practice of surrogacy. For instance, there are partial bans in Brazil, Israel and the UK, but no regulation at all in India, Finland, Belgium, and Greece (Pande, 2009: 381). The international community has thus failed to set standards regarding how these new reproductive opportunities need to be understood. Wilkinson (2006: 137) makes a distinction between *partial* or *straight* surrogacy, whereby the surrogate mother is the genetic mother, and *full* or *host* surrogacy, whereby the "surrogate provides only gestational services and the gametes are provided by others (often, but not necessarily, the commissioning parents)". Both forms of surrogacy are controversial; the first for making the baby a commodity that can be given away under certain legislations and the second because the surrogate mother is treated as a reproductive machine to be rented by the reproductive cycle. Teman's (2010) ethnographic study of surrogacy in Israel, a country relatively liberal in terms of allowing surrogacy and where the demographic policy emphasizes the need for new babies, ultimately embedded in the idea of a unified Jewish people, suggests that most surrogate mothers were able to balance their various roles and expectations, both practically and emotionally. A few of the women interviewed claimed they had to defend their bodies against worried parents, asserting that they themselves decided what to eat and what to do during pregnancy. In the Israeli case, surrogacy was accomplished within a relatively homogeneous group of people; however, in many cases, surrogacy is provided by poor women in, for instance, India, leading to the inevitable concerns about new and ethically questionable forms of neocolonialist exploitation. Pande's (2009) study of commercial gestational surrogacy in Anand, Gujarat shows that the gift economy enacted in other domains of reproductive medicine may not apply to the Indian case because the monetary compensation overshadows the entire relationship between the surrogate mother and the parents, making the apparent economic inequalities a major issue to handle prior to any full-scale recognition of what has been called "reproductive outsourcing", an overtly cynical term pointing to the opportunities for wealthy Westerners to make use of poor Indian women as gestational resources. In Anand, Gujarat, average income is INR 2,500 (USD 60) per month, and a surrogacy generates USD 3,000–5,000, "equivalent to nearly ten years of family income" (Pande, 2009: 383). The women in Anand recruiting new surrogate mothers earn INR 10,000 (USD 300), which is a substantial sum in rural areas of India. As Pande (2009: 391) remarks, "surrogacy upsets the traditional moral framework in which reproduction is regarded as a 'natural fact' grounded in love, marriage and sexual intercourse … replacing it with a commodification of bodies, feelings, and values." However, the Anand women deal with such inconsistencies by claiming that they are working hard to carry the baby and thus deserve the pay they receive: "The child is a product of its (gestational) mother's sweat and blood, a fruit of the labor and effort of gestation and this confers identity to the child" (Pande, 2009: 387). Despite reports regarding the ostracism and stigmatization of surrogate mothers, the pay received for their clinical labor is too high for these poor women and their families to resist. As a consequence, there is an apparent economic rationale in favor of surrogacy as one adequate method in reproductive medicine, but there is also the more lingering concern about whether or not everything can legitimately be bought for money. There are Western families endowed with the economic resources needed and women willing to provide gestational surrogacy, but can the exchanges between these two parties be separated from the wider social and institutional fabric? Thomas Frank (2008) conceives of gestational surrogacy as the ultimate triumph of neoliberal thinking, turning virtually any conceivable good and service into a commodity, leading to a corruption of the sense of worth and value and constituting little more than a loss of respect for human life. On the other hand, Cohen and Athavaley (2009) tell another story of an Arizona couple who became parents after using surrogacy as their last resort after spending long periods of trying in every possible way to become parents.

No matter how one regards the emerging practice of using surrogacy, reproductive medicine, no matter how advanced it becomes, still relies on the gift relations between human beings. The "gift of life" may be accompanied by monetary compensation for the ordeal on the part of the donor but it is still a procedure embedded in the willingness to help the other, the childless couple or single woman so anxious to become parents on their own.

When gifts become illegitimate: Blurred boundaries in medical research funding

While reciprocity is commonly portrayed as a generic, even primordial mechanism for structuring human societies, there are some social domains where reciprocity becomes problematic. The relationship between academic medicine and the pharmaceutical industry is one such area riddled with concerns (Brody, 2007). Lakoff (2006: 140) argues that the "ubiquity of gifts" from the pharmaceutical industry to doctors and researchers has recently been highlighted as a case of potential "conflict of interest". The line of demarcation, between "rational pharmacology" and "drug promotion", needs to be maintained in order to uphold the prestige and credibility of either partner and is compromised by an all-too-close relationship between the producers of knowledge and the commercial division. Still, as Lakoff (2006: 140) notes, "marketing and expertise cannot be so easily distinguished: pharmaceutical companies are producers not only of pills but also of knowledge about their safety and efficacy, and their gifts to doctors of travel to conferences and workshops provide access to the latest expertise. The fortress that is supposed to guard against the crude logic of profit – biomedical expertise – is itself ensconced in the market." Fishman (2004), who studies the development of so-called sexuopharmaceuticals, e.g., drugs dealing with what has been called female sexual disorders, points to the complex ecology of the field of medicine:

> Pharmaceutical companies are not alone in standing to gain from such [mutually beneficial] arrangements and performances [collaboration with academic researchers]. The researchers themselves, in addition to receiving financial rewards as consultants for pharmaceutical companies, also gain professional recognition, funds for their research departments and laboratories, publications and, often, media attention through related public and professional activities. (Fishman, 2004: 188)

The marketing of drugs is largely dependent on the prestige and credibility of scientific communities, and scientific claims regarding efficacy and therapeutic benefit make a substantial difference to how well the drug is received in the marketplace. Even though there is detailed regulatory control of how scientific work is funded and conducted, there are still examples of violations of the scientific ideology of free-standing researchers. Healy (2006) has called attention to the use of ghost-writing in the pharmaceutical industry whereby companies hire freelance writers to pen papers that are signed by leading scientists and then submitted to scientific journals. The scientific ideology that emphasizes that researchers should craft their own papers in order to be fully accountable for their positions is, in so doing, bypassed as specialized writers produce texts that stress the research findings benefitting the sponsor. "[G]hostwriting is no longer occurring only in peripheral journals and affecting only review articles. It happens in the most prestigious journals in therapeutics, and it probably happens preferentially for papers reporting randomized trials and other data-driven papers", writes Healy (2006: 72). In addition to the morally-questionable practice of ghostwriting, the term "honorary authors" is used when leading scientists in a field are listed as co-authors of papers despite having made little contribution to the writing of the paper. Mirowski and van Horn (2005: 528) make reference to bibliometric studies of journal articles published in the field of medicine: "In the aggregate, 19% of the papers had evidence of honorary authors, 11% had evidence of ghost authors, and 2% seemed to possess both ... [in another study] 39% of the reviews had evidence of honorary authorship, while 9% had evidence of ghost authorship." In these cases, the reciprocal relations between scientists and pharmaceutical companies become illegitimate as the third party, the general public, fails to see the full complexity of exchanges of money and credibility across institutional boundaries. Scientific ideologies and standards for practice, enacted and fine-tuned over decades or centuries of work, are violated; in the end, both the scientists and the pharmaceutical companies are undermining their own positions.

Reciprocity and innovation

Gift economies and innovative capacities

Ferrary (2003) studied the computer industry in Silicon Valley in the San Francisco Bay Area and speaks of "gift exchange" as an important social mechanism regulating the renowned computer cluster. At the center of these social relations are the venture capital firms which serve a manifold of functions in distributing resources, connecting various actors and competencies, and maintaining control of the industry (Ferrary, 2003: 122). Even though access to informed and experienced venture capitalists is one of the single most important factors explaining the dominance of Silicon Valley, the cluster is a sophisticated ecology that accommodates a variety of experts and specialists who, in various ways, have developed mutual dependencies and intricate webs of social relations. "Silicon Valley should not be understood as simply an accumulation of resources, but as a multitude of social networks that assure an optimal diffusion of information between complementary economic agents", suggests Ferrary (2003: 121). What propels the computer industry is creativity and information, the capacity of certain individuals to conceive of new and innovative ideas about how to exploit consumer demands and the available sources of information. Since neither creativity nor information easily lend themselves to market transactions, the circulation of information is located in reciprocal relations between key actors within the industry: "Because information itself is not marketable, all given information establishes credit for the provider, which can only be erased by a counter-gift", suggests Ferrary (2003: 131). The reciprocal relations structured around the gift/counter-gift exchange thus institute a new temporal order within the industry. One very prestigious venture capital firm may be unable to handle all incoming requests for funding and, rather than rejecting a business proposal, it passes this on to one of the less prestigious, but still credible, venture capital firms. The entrepreneur may then receive his venture capital at the same time as the first venture capital firm is giving the "gift" of the business proposal to the second venture capital firm. In return, the first venture capital firm may, some time in the future, or sooner or later, ask for a favor, a counter-gift from the second venture capital firm, be it help in evaluating a business proposal, a suggestion for a suitable board member for a company, or virtually any service that the company is capable of providing. Such a counter-gift closes the gift-exchange cycle and the relationship between the two firms is maintained over time. Gifts given now may be repaid in a relatively distant future but, unless the gift is repaid, there will be no possibilities of escaping this obligation. However, there is always, as Ferrary (2003: 131) notes, the possibility of refusing the gift, but this entails avoiding the cycle of exchange altogether. At the same time, if an actor wants to prosper on the basis of gaining access to the circulation of know-how and information, he/she will not be able to stay outside the gift economy of Silicon Valley for very long. Ferrary (2003) suggests that the gift economy is not a strictly economic matter; on the contrary, it is something that needs to be understood within the full scope of human and social interaction, including both economic and non-economic exchanges. For Ferrary (2003), the dominance of Silicon Valley in the computer industry needs to be explained on the basis not only of economic conditions, but also of social and cultural traits and traditions (see also Saxenian, 1994).

Various Internet communities, creating and sharing data and information, is another field characterized by innovative capacities that draw on instituted gift economies. The economic value generated is bound up with the circulation of know-how and expertise. Currah (2007: 469) uses the term "electronically mediated gift economies" to capture the idiosyncratic Internet culture. In contrast to the anthropological view of gift economies, which is relatively geographically narrow and in many cases relies on face-to-face interaction, the Internet gift economies are, says Currah (2007: 469), "usually vibrant and geographically extensive in their structure and operation": "This is especially the case they are compared by traditional precedents, which tend to be organized around the exchange of physical objects in locally defined communities, bound together by tightly woven definition of reciprocity and trust." Internet gift economies are, on the contrary, largely anonymous and based on the various peer-to-peer (p2p) software programs designed to enable the almost effortless sharing of data.

> Unlike traditional Maussian procedures, Internet gift economies do not explicitly depend upon dynamic moral or reciprocal obligations between individual participants. To be sure, the networked environment does enable exchanges between friends and family (through so-called

'strong' social ties), but, in general, the networked environment enables – and indeed, actually flourishes from – the exchange of resources
between millions of strangers (through 'weak' ties). (Currah, 2007: 474)

In addition, while most anthropological theories of gift economies presuppose a scarcity of resources to share, making the exchange of gifts not only ritual and ceremonial in character but also part of an economy wherein different groups (clans, tribes, even peoples) can specialize in, for instance, agriculture or fishing, the Internet needs to be understood on the basis of an "economics of abundance" (Currah, 2007: 476), i.e. a situation whereby data and information can instantly be reproduced and/or modified due to digital media enabling such operations (Manovich, 2001). Therefore, the agents of the gift economy on the Internet are not so much concerned with securing foodstuffs or other necessary resources as they are with aiming to accumulate prestige and status in the eyes of the virtual community: "Quite simply, the overriding objective in a gift economy is to *give away* resources to secure and retain status", says Currah (2007: 475) (see also Rehn, 2001). The study reported by Bergquist and Ljungberg (2001), on the gift economy of open source communities, supports Currah's (2007) arguments, pointing out that the status of the community is largely a matter of how much attention is received from peers when making a modification to the shared code. By giving away his/her expertise in computer programming to the community, the individual actor receives attention and prestige as a counter-gift; consequently, there is a certain hierarchy based on the contributions made and how these contributions are received. The open source community thus operates on the basis of trading in programming expertise, and the credit and recognition generated within the community.

Social capital and innovation

Another key concept in the social science literature as well as the organization theory corpus, pertaining to reciprocity, is the concept of social capital (Maurer and Ebers, 2007). Social capital has been defined in many ways but is commonly defined as shared norms and beliefs instituting trust and mutual recognition on the part of social actors. Coleman (1988) locates social capital to the social structure of any social formation:

> Social capital is defined by its function. It is not a single entity but a
> variety of different entities, with two elements in common: they all
> consist of some aspect of social structures, and they facilitate certain
> actions of actors – whether persons or corporate actors – within the
> structure. Like other forms of capital, social capital is productive, mak-
> ing possible the achievement of certain ends that in its absence would

> not be possible ... Unlike other forms of capital, social capital inheres
> in the structure of relations between actors and among actors. (Cole-
> man, 1988: S98)

Social capital is, thus, the mechanism enabling and forming relations "between and among actors". Portes (1998: 7) adheres to such a view of social capital: "Social capital inheres in the structure of their relationships. To possess social capital, a person must be related to others, and it is those others, not himself, who are the actual source of his or her advantage." Subramaniam and Youndt (2005: 451) define social capital less abstractly as "[t]he knowledge embedded within, available through, and utilized by interactions among individuals and their networks of interrelationships". Social capital is also commonly separated from *human capital*, i.e. the individual competencies or skills: "Just like physical capital is created by changes in materials to form tools that facilitate production, human capital is created by changes in persons that bring about skills and capabilities that make them able to act in new ways", argues Coleman (1988: S100). Human capital is, then, "intangible" but "embodied in skills and knowledge acquired by an individual" (Coleman, 1988: S100–101). "[S]ocial capital is a quality created between people, whereas human capital is a quality of individuals", as Burt (1997: 339) straightforwardly puts it. Subramaniam and Youndt (2005: 451) say that human capital is "[t]he knowledge, skills, and abilities residing within and utilized by individuals".

The relationship between social and human capital is of key interest to both sociologists and organization theorists inasmuch as the former helps to leverage investment in the latter. Subramaniam and Youndt (2005: 459) thus strongly emphasize the role of social capital:

> To effectively leverage investments in human capital, it may be imperative for organizations to invest in the development of social capital to provide the necessary conduits for their core knowledge workers to network and share their expertise. Organizations that neglect the social side of individual skills and inputs and do not create synergies between their human and social capital are unlikely to realize the potential of their employees to enhance organizational innovative capabilities. (Subramaniam and Youndt, 2005: 459)

Social capital, defined as the norms prescribing how, for example, the human capital of firms or organizations should be jointly exploited, thus serves to enable the circulation and distribution of human capital. As Portes (1998: 15) emphasizes, the research on social capital "strongly emphasizes its positive consequences"; however, he also remarks that it is a "sociological bias" to see "good things emerging out of sociability" while "bad things are more commonly associated with the behavior of homo oeconomicus". In other words, for sociologists, social capital is not a value-neutral concept but denotes, on the contrary, the instituted norms and beliefs helping to further reinforce the sociality of human beings. A more moderate view, like that of Subramaniam and Youndt (2005), would be that individual and collective investments in human capital benefits from having access to social capital but social capital without human capital is relatively empty, suggesting that organizations cannot expect much from their investment in human capital unless it is accompanied by social capital. Nahapiet and Ghosal (1998: 243) make a distinction here between the *structural*, the *relational*, and the *cognitive* dimensions of social capital whereby the structural dimension denotes the formal rules and regulations prescribing how know-how should be used in an organization, the relational social capital dimension directs the relationship between collaborating individuals, and the cognitive social capital dimension is the shared modes of thinking that structure everyday work. Nahapiet and Ghosal (1998) also emphasize that social capital supports the exploitation of what they call "intellectual capital" – "the knowledge and knowing capability of a social community, such as an organization, intellectual community, or professional practice" (Nahapiet and Ghosal, 1998: 245). In contrast to other scholars, Burt (1997) speaks of social capital as an individual asset, some kind of general social competence that helps to achieve recognition of human capital: "Managers with more social capital get higher returns to their human capital because they are positioned to identify and develop more rewarding opportunities", suggests Burt (1997: 339). Individuals having these social skills, Burt (1997: 49–50) demonstrates, get promoted earlier and get more bonuses. In addition, the more marginal the manager is vis-à-vis the organization – what Burt (1997: 355) calls *boundary managers* – the higher the value of the social capital.

Notwithstanding Burt's (1997) contributions, the common view in the literature of social capital as a collective resource, a form of infrastructure for the exploitation of human capital, is more fruitful than thinking of social capital as some kind of general "social competence" or "people skills". Social capital is, then, in the structural view of Coleman (1988) and others, which emphasizes the reciprocal relations between social and human capital, agency and structure. Consistent with the writing on social structure, social capital is insignificant unless there is human capital in place that can be put to use; as soon as human capital is acquired, it needs to be coordinated using existing social capital. Social capital is embedded in the principle of reciprocity, the principle that human beings must demonstrate at least the minimum amount of trust and mutual recognition in order to be able to exploit their human capital effectively. This does not suggest, however, that access to social capital, of necessity and under all circumstances, leads to benefits for the organization. Bresnen et al. (2005: 237) suggest that one problem with social capital is that it "[c]an be a very costly and inefficient means of obtaining information as it involves considerable effort in establishing and maintaining relationships". In addition, these scholars also argue that "within-group cohesion" promotes an "inward-looking perspective" that creates problems when there is a "need to access new information or to harness different expertise to address novel problems" (Bresnen *et al.*, 2005: 242). In short, reliance on social capital as a mediator of know-how establishes a certain conservatism that leads to cognitive lock-in effects.

Regardless of such criticism, studies of innovative capacities demonstrate that social capital and a social orientation are a *sine qua non* for successful innovation activities. Grant and Berry (2011) explore the relationship between creativity, an important driver of all innovation, and what they call "prosocial motivation", i.e. the capacity to take on the role of the other during innovation work (see also Leonardi, 2011). In their review of the extensive literature on intrinsic motivation and creativity, Grant and Berry (2011: 77) found that "intrinsic motivation cultivates a primary focus on novelty but not necessarily on usefulness". They thus propose that "other-directed psychological processes" play an important role in "guiding employees towards considering ideas that are not only novel, but also useful". That is, intrinsic motivation is helpful in promoting creative but also useful ideas. At the same time, "[o]ther-focused psychological processes can constrain creativity by fostering a focus on conformity, which reduces employees' capabilities and motivation to think divergently" (Grant and Berry, 2011: 91). Such a strong emphasis on intrinsic motivation is, thus, suggest Grant and Berry (2011), misleading. Based on their findings, they propose some managerial implications:

> Managers typically seek to stimulate creativity by creating conditions that are conducive to intrinsic motivation, such as designing challenging and complex tasks, providing autonomy, and developing supportive feedback and evaluation systems. Our research suggests that these practices run the risk of enhancing intrinsic motivation without also cultivating the prosocial motivation and perspective taking that can facilitate the production of ideas that are creative in context. (Grant and Berry, 2011: 93)

If creativity and usefulness are, as Grant and Berry (2011) suggest, two complementary terms, then the ability to create relationships with end-users and other significant stakeholders may further enhance a firm's innovative capacities.

The literature on scientific communities, or more specifically what Håkansson (2010) calls, with reference to Knorr Cetina (1999), *epistemic communities,* strongly emphasizes the institutionalized social order whereby social, organizational, and human capital are intimately bound up and interrelated. Within epistemic communities, says Håkansson (2010: 1805), "[i]ndividuals often possess the same, experiential knowledge and can with relative ease pass it on in personal contacts, and with some investment it can usually be articulated within shared symbolic and theoretical frames." As a consequence, a member of an epistemic community thinks, perceives, and acts in accordance with the instituted professional practices. Ludwik Fleck (1979) addresses epistemic communities as *Denkenkollektiv*, "thought collectives", sharing a specific *Denkenstile*, "style of thought", that is shared across spaces and local communities. A series of studies of, for instance, bioengineers (Sharp, 2011), meteorologists (Fine, 2007), nanotechnology scientists (Selin, 2007), protein crystallographers (Myers, 2008), oceanographers (Goodwin, 1995), zoologists, (Roth and Bowen, 1999), cattle breeders (Grasseni, 2004), radiologists (Alac, 2008), biologists (Rheinberger, 1997), cytologists (Keller, 1983), particle physicists (Traweek, 1988), computational chemists (Torgersen, 2009), and surgeons (Huckman and Pisano (2006) has demonstrated that professionalism lies precisely in one's capacity to apply disciplined expertise while simultaneously being able to critically reflect on one's professional work. Studies of tertiary professional training and the work of neophytes in, for example, architecture (Lymer, 2009), law and business school graduate students (Schleef, 2006), medical school graduate students (Becker et al., 1961), surgeons and residents (Kellogg, 2009; Pratt, Rockmann and Kaufman, 2006), nurses (Benner, 1984), and accountants (Anderson-Gough, Grey, and Robson, 1998) indicate that, not only do newcomers to a professional field learn to master their day-to-day routine work, they also have to be able to question their own role and the predominant practices. The skilled professional is thus capable of striking a delicate balance between, and adhering to, the instituted order of practice and activities seeking to modify and adjust the profession to emerging conditions and new social needs.

Summary and conclusion

Competitive capitalism is commonly associated with a preoccupation with personal gain and the short-sighted concern regarding profit. In this master narrative, the self-interest of the economic agents serves, as Adam Smith famously puts it, as the "invisible hand" that guides manifold economic activities. While such an image of self-interested and largely autonomous agents is partly capable of explaining economic agency, it also needs to be complemented by collectivist accomplishments. For instance, in the medieval mercantile economy, the guilds served the role of protecting the individual's interests (Epstein, 1998; Krause, 1996). During the contemporary period, the professional communities and stakeholder organizations (e.g., trade unions) serve similar roles. Such communities and organizations are not only based on monetary exchanges and transactions but are also inherently social, relying on the intellectual and emotional support provided by peers, colleagues, and members of the same professional community. This intellectual and emotional support is embedded in reciprocity, in gift economies, wherein actors jointly constitute authority and expertise on the basis of collectively affirming the skills of individuals. Social capital, reputation, and expertise are some of the socially-grounded competencies that are embedded in reciprocal relations. In addition to such "abstract" forms of reciprocity, there is also a very tangible and material exchanging of gifts in the contemporary capitalist regime. Goods and services are exchanged as part of an overarching economic trade; consequently, one must not be too quick to separate economic and social relations.

However, due to laws and regulations and norms regarding conduct, there is evidence of illegitimate reciprocity, with bribery, nepotism, and the conditional sponsoring of scientific work being some of the concerns deriving from reciprocal relations between actors which have destructive effects on other social relations (e.g., the illegitimate reciprocity between the pharmaceutical industry and the physician may negatively impact the relationship between the physician and the patient). Still, in order to accomplish certain objectives, economic agents give away things – ideas, know-how, commodities, credibility, and so forth – in order to receive something in return later on. That is, rather than being strictly based on scarcity, economic action is based on abundance, on what can be passed on today for future benefit. By giving away various kinds of gifts – tangible or abstract or immaterial – social relations are sustained over geographical distances and over time, and future economic transactions may be safeguarded. The strictly self-interested and calculative economic agent gradually loses his/her legitimacy, if he/she refuses to be part of the gift economy; the economic agent may choose to only receive and never to give, but such a strategy will lead to a subordinate position in the field. Refusing to participate altogether will eventually cut the economic agent loose from the field. Thus, the paradox of "competitive capitalism" is that, in order to make the economy grow, the economic agent has to give away economically-valuable resources; only by sharing resources can new resources be accumulated and acquired.

The reliance on gift economies does not suggest that competitive capitalism entails the democratic sharing of resources. On the contrary, it underlines the fact that socially-relevant actors will benefit from their positions within the gift economy (e.g., the venture capitalists of Silicon Valley), while other actors will receive little or almost nothing. The concept of the gift economy does not, then, suggest the mindless squandering of resources; instead, all kinds of gifts serve to establish and sustain social relations that may be used to accomplish economic transactions later on. Thus, the concept of "gift giving" needs to be understood, not as an act of generosity, but rather as an institutional process demanding its counter-gift. Just like capital in general, the gift economy is *circulatory*, operating on the basis of the ongoing exchange and flow of gifts.

Chapter 4.

Squandering

Exordium

In the study of assisted fertilization, introduced in the preceding chapter, one of the concerns for the actors in the field was how to balance technoscientific and medical opportunities with the regulatory framework setting the boundaries. "The regulatory framework by definition lags behind the scientific advancement in the industry", argued one of the interviewees. One concern was that, while there was an endemic shortage of egg donors, all of the clinics were holding a large quantity of embryos in their freezers. Under normal conditions, and in the case of successful assisted fertilization, a number of eggs are retrieved from the female patient, with a few of these eggs being fertilized and grown in various media over a period of approximately forty-eight hours before the embryo is transferred to the uterus. When the embryo develops into a baby, the patient no longer needs the other embryos and, consequently, there will be a surplus of embryos while there is a shortage of eggs. "Sperm donation was not permitted until the mid-1980s, egg donation not until 2003. Embryo donation is not legal in Sweden in any way, shape, or form, or surrogacy for that matter", remarked one of the professors in reproductive medicine, testifying to how the legal framework prohibited what the interviewees referred to as "embryo donation" or even "embryo adoption". The professor said that this biological resource is wasted when the storage period comes to an end, arguing that this is an unfortunate situation for reproductive medicine and for patients who fail to get pregnant:

> We have tens of thousands of embryos in freezers here in Sweden but we mustn't keep them any longer than five years. What can we do with them after that? Is it better to throw them away than to use them in embryonic stem cell research? We believe it's better to use them. And we do have legislation which permits that, with the consent of the patient. (Professor of reproductive medicine)

Another interviewee, a gynecologist, also stressed the need to oversee the legal framework that renders thousands of embryos waste products from successful assisted fertilization cycles:

> Q: This issue of donating embryos?
>
> A: Well, we want that to be legalized. Right now, we keep all these frozen embryos that we will be throwing away after five years … What you may call an adoption, adopting embryos is much better than adopting small children that have already been born. (Gynecologist, founder of private assisted fertilization clinic)

Natural and biological systems are characterized by overflow and excess and the regulatory framework seeking to sort out what can legitimately be done, for instance, the field of reproductive medicine is always at risk of lagging behind, failing to take into account what could be considered ethical, morally sound, and generally socially acceptable in the domain of assisted human reproduction – on the one hand, the excess of biological reproductive materials and adequate technoscientific expertise; one the other, lawmakers and regulators, politicians, scientists and lawyers, in charge of formulating and enacting laws and regulatory frameworks that balance various objectives and resources. The taming and controlling of nature by means of legal texts often underrate the excess and squandering of resources in all sorts of biological processes (see, for example, Calvert, 2007). Wasting and squandering lie at the very heart of biological creation.

Introduction

> "A great civilization is first and foremost a civilization that has a waste-disposal system. So long as we do not take that as our starting point, we will not be able to say anything serious."
>
> Jacques Lacan (2008: 65)

"The modern industrial city, the rise of the nation state, and even the struggle for clean and proper language, are all inextricably bound in a struggle with sewage and waste; the Victorian engineering of sewers and pipes, subterranean gutters and drains, aqueducts and the viaducts that we see around our suburbs and towns, all components of a complex infrastructure required to maintain the circulation of water so vital to the body and its city", remark Rehn and O'Doherty (2007: 102), contributing to a narrative on infrastructure, all too often ignored by the social sciences (Star, 1999. On the underground of cities and societies more broadly, see Williams, 1990). Contemporary society, at least in the Western world, overflows with resources. No other historical society has ever been able to produce such vast amount of food, goods, and commodities (on the history of ideas regarding waste, see Gee, 2010). The increase in efficiency in agriculture and manufacturing has been remarkable and today it is nature *per se*, not the human capacity to transform it into commodities, that constitutes the limit to how much humans can consume. McCloskey (2006: 16) notes that, between 1800 and 2006, the period of capitalist accumulation, "the amount of goods and services produced and consumed by the average person on the planet [had] *risen* since 1800 by a factor of about eight and a half" (McCloskey, 2006: 16). At the same time, the economic output of the capitalist economy has grown at an even higher rate between 1820 and 2003: "In 1820 … the total output of goods and services in the capitalist world economy was worth $694 billion (in 1990 constant dollars), By 1913 it had risen to $2.7 trillion; by 1950, it was £5.3 trillion; in 1873 it stood at $16 trillion; and by 2003 nearly $41 trillion" (Harvey, 2010: 27). Such statistics suggest that there has been close to a 60–fold increase in output in constant dollars (taking 1990 as the base year). "Throughout the history of capitalism, the actual compound rate of growth has been close to 2.25 per cent per annum", reports Harvey (2010: 27). By any stretch of the imagination, we are living in what John Kenneth Galbraith referred to as the "affluent society". "A hundred years ago, households in OECD countries spent 80% of their income on the basic needs of food, clothing, and housing. This figure has dropped to 30–40% today" (Beckert, 2011: 106)[23]. The distribution of all the wealth produced globally is quite another matter, both within the Western world and between regions and nations, in particular. In his *Stone Age Economics* (1972, Chapter 1, reprinted in Sahlins, 2000), anthropologist Marshall Sahlins (2000: 95) argues that the original affluent society is not in fact a society of plenty like ours, but one "[i]n which all people's material wants are easily satisfied". "The world's most primitive people have few possessions *but they are not poor*", says Sahlins (2000: 133), on the basis of his definition of poverty as "a relation between people" and a "social status" (Sahlins, 2000: 133). Sahlins thus suggests that poverty is the immediate effect of a society wherein enormous amounts of resources are accumulated and unevenly distributed; in a

---

23 It is noteworthy that such statistics tend to conceal the uneven distribution of wealth in the contemporary economy. The loss of relatively well-paid jobs in manufacturing in, for instance, the UK and the growth of service jobs (around one million people work in call centers in the UK today), paid roughly a quarter less (GBP 24,343 and GBP 20,000 average annual wage, respectively – female hairdressers, one of the worst paid jobs in Britain today, make as little as GBP 12,000 a year) were compensated for by an inflow of capital (Jones, 2011: 151–152). As a consequence, a lot of consumption during recent decades has generated debt: "In 1980 the ratio between debt and income was 45. By 1997 it had doubled, before reaching an astonishing 157.4 on the eve of the credit crunch in 2007. As people's purchasing power slowed, more and more credit was splashed out on consumer goods. Between 2000 and 2007, consumers spent £55 billion more than their pay packets, courtesy of the plastic in their wallet or hefty bank loans" (Jones, 2011: 158). In other words, while affluent social strata may benefit from larger household budgets, poverty is increasing in the traditional working class neighborhoods and regions.

society where few resources are accumulated for sharing, there is, by definition, no poverty because there are no possibilities of discriminating between social groups and individuals on the basis of their access to resources. Sahlins seeks to de-familiarize the very idea of an affluent society, saying that economic resources are not only a blessing but also something that leads to questions of justice, a question that the existing economic regime has been most unsuccessful in solving. In Sahlin's view, "the original affluent society" is something entirely different from our own society – a leisure society rather than a society of harsh competition. One should be wary of the risks of expressing overt nostalgia regarding past and obsolete social formations, but Sahlins argument points to some of the difficulties facing the present economic regime, not only the producing of wealth but also, *ipso facto*, poverty (see, for instance, Davis, 2006).

In this chapter, the concept of squandering, the wasteful consumption of the excess, and the overflow of resources produced in all forms of societies, is examined. The French social philosopher Georges Bataille argues, in his three volume monograph *The Accursed Share*, that all historical societies have instituted some mechanisms for the wasting of excess resources.[24] In ritual and ceremony, such seemingly irrational consumption of what Bataille calls the *accursed share*, the part of the crop or the production that is dispensable, that can be wasted, is conspicuously squandered. Bataille argues that it is precisely such rituals, whereby what has been accumulated over time and through hard work is wasted, that make humans human, demonstrating to the members of a particular society that they belong to a social formation that is capable of slowly accumulating resources but also of quickly consuming them. This general anthropological and historical observation is by no means undisputed and Bataille is, in many ways,

---

24 For Marxist economists and social theorists, the accumulation of surplus capital in society is not a trivial or marginal concern but lies at the very heart of the capitalist regime of accumulation and regulation. Harvey (2010: 26) speaks of "the capital surplus absorption problem" whereby the surplus capital generated in increasingly efficient industries needs to find new "profitable outlets". In the early years of the new millennium, prior to the start of the 2008–2009 economic recession (for an overview, see Stiglitz, 2010a, b), the financial markets were one such outlet, making the turnover of the derivatives market larger than world economic output for the first time ever in 2003 (Harvey, 2010: 23). Between 2003 and 2007, the turnover of the derivatives market grew faster than world economic output; "The global market for derivatives rose from $41 trillion to $677 trillion in 1997–2007", adds Callinicos (2010: 74) (see Mackenzie [2011] for a discussion on the various derivate instruments and packages of derivate instruments developed in the finance industry and the difficulties associated with assessing their risks). In addition, while US corporate profits in manufacturing have been in decline since the early 1950s (Harvey, 2010: 22), the finance sector is enjoying profit growth during the same period. In the latter half of the 1990s, accumulated profits in the finance sector exceeded those of manufacturing industry for the first time. Harvey (2010) thus suggests that surplus capital has been channeled into the finance sector, pumping capital into, for instance, the real estate market, in turn leading to an unprecedented level of household debt. It is noteworthy that liberal economists adhering to a monetarist doctrine tell another story, pointing to the migration of capital into markets that have

a difficult social theorist in terms of advancing ideas that may not be accompanied by robust empirical evidence (Hegarty, 2000; Noys, 2000). The concepts of squandering and waste are thus used in the broadest possible sense of the term in this chapter and in this setting, as the proclivity of contemporary society – in competitive capitalism – for accumulating resources which, to some extent, are wasted, i.e., not consumed under the same monitored conditions as they are produced. This means that there is a constant endemic overproduction of resources – material, emotional, visual, cognitive – that are never used, never brought forward. One may easily think of such squandering of resources, in moralistic terms, as a violation of norms and even taboos. However, if one examines nature, it too is characterized by overflow and waste. For instance, the human genome, surprisingly limited in terms of discriminating a human from a cow, a bee, or a dandelion, is made up of between 95–97 per cent so-called "junk DNA", DNA that seemingly does not have a clear role and function in producing the amino-acids and proteins that constitute the organism (Kay, 2000: 2; Rose, 2007: 270, note 7). Such observations have led to renewed interest in Georges Bataille's thinking in the scholarly literature that addresses the life sciences. Thacker (2006: 130), for instance, makes a reference to Bataille in his study of genomics research:

> In a sense, despite the biotech industry's effort to manage the genome, the genome itself (like 'life itself') appears to be nothing but excess information, with all the contradictions that the phrase implies. Bataille proposes that an organism always produces more than it needs and that, from a systems-wide ecological perspective, abundance, not scarcity, is the rule of life. (Thacker, 2006: 130)

In the same vein, Sunder Rajan (2006: 113) uses Bataille to understand the func-

undervalued some of the resources, e.g., the housing market. Despite such arguments, many commentators argue that that the 2008–2009 recession was largely caused by the widespread issuing of sub-prime loans in the US, propelled by short-sighted bonus systems and systematically overrating access to capital over the economic cycle among low-income groups, thus exposing the finance sector to excessive risks (Davis, 2009a, b). Despite the tragic cases of families being evicted from their homes after defaulting on their loans, many governments, both in North America and Europe, chose to bail out the banks with taxpayers' money in order to safeguard a stable supply of capital, thus orchestrating an unprecedented transfer of taxpayers' money from the public sector to the finance sector. The bank bailout in 2008–2009 was estimated by the IMF to have cost 12.7 per cent of GNP in the US and 9.1 per cent in the UK (Callinicos, 2010: 88). In Ireland, for instance, which faced a situation in the fall of 2010 whereby many of the banks were on the verge of collapse due to excessive economic risk-taking, the Dáil passed a budget that included extensive welfare-state spending cuts, reduced public-sector wages, and a series of other withdrawals of public benefits and services. The economic and financial crisis of 2008–2009, and its aftermath, further accentuated the critique of the neoliberal agenda being driven by institutions such as the IMF and the WTO (Chorev and Babb, 2009). In the spring of 2009, during the G20 meeting, at the height of the crisis, British prime minister Gordon Brown declared the death of the so-called "Washington consensus", the "famous list of market-liberalizing policy descriptions that [had] guided the previous 20 years of economic policy" (Chorev and Babb, 2009: 459).

tioning of the life sciences. Perhaps "waste" is not an adequate term to denote scientific work because waste has moralistic overtones of the meaningless destruction of accumulated resources. In the life sciences, waste is, nevertheless, the rule rather than the exception. Says one life science industry interviewee, cited in Petryna (2009: 120): "Of 10,000 chemical compounds only 100 are tested on animals. Of those 100, 10 go into human testing and 1 becomes a marketable drug. The 9,999 that fall by the wayside have cost money somewhere. A lot of money is wasted." In order to win, one must bet, and betting in, for example, the pharmaceutical industry is quite costly since the cost of developing a new drug is astronomical: "The cost of generating a single approved medicine is claimed to be over $600million," reports Barry (2005: 58).

No matter what concept or term is used to denote the overflow of resources and their purposefully wasteful handling, excess is one of the marks of the contemporary economy. The dismal science of economics is based on what Doel (2009: 1056) calls "miserly thinking", whereby one is "fated to allocate scarce resources, obligated to utilize deficient means, and duty bound to minimize waste". Such thinking largely fails to recognize the "[f]undamental importance of excess to both the maintenance and the transformation of social systems and spatial structures" (Doel, 2009: 1057). While conventional economic theory is "one of parsimony, calculation and utility – a conception which remains the mindset of most business practitioners", suggest Rehn and O'Doherty (2007: 100), Bataille's concept of the general economy and the concept of excess – "that which is above and beyond the bare necessity and its barren land of utility and harsh Puritanism" (Rehn and O'Doherty, 2007: 103) – serves to enable entirely different images of the economy. "What Bataille calls the general economy", suggests Doel (2009: 1057), "entails the restoration of baselessness, purposelessness, and superfluousness to the world". Bataille thus introduces an "unusual and bizarre ontology" (Rehn and O'Doherty, 2007: 101), one that is at odds with conventional understandings of economic reason. Rather than assuming that all resources generated in a society are mindfully put to use, carefully managed, and rationally portioned, it is possible to account for an alternative trajectory. "Work is not inherently of value", says Doel (2009: 1057): "Whether the expenditure of energy is of value – in whatever shape or form (e.g., as a use-value, a sign-value, or an exchange-value) – depends upon the ends to which it is put." In the economic thinking advanced by Bataille, utility is not the starting point for the analysis but is, in itself, an outcome of economic reasoning, a social accomplishment in its own right. Economic life is concerned with both production and expenditure.

In this chapter, the principle of squandering, expenditure, and waste will be discussed as something that complements the principles of playfulness and reciprocity as something that lies at the very heart of economic agency and situated rationalities. As in the previous chapters, there are intersections and connections between the three principles. Playfulness is part of the squandering and play itself, as Caillois (2001a) put it, is a form of wasting time and energy. Squandering also rests on the principle of reciprocity inasmuch as it would be meaningless to waste unless this testifies or represents the strength and vitality of the society producing and wasting the resources. On the other hand, there is no reciprocity without at least a minimum amount of waste, of unnecessary rituals and undertakings, orchestrated to gratify and affirm friendship or collegiality. However, throughout this chapter, the concept of squandering is used in less anthropological terms than those potentially intended by Bataille, advancing the ritual waste of resources in historical societies as a form of critique of the obsession with utility during the contemporary period. In this context, squandering is less ceremonial and less affirmative of the underlying social formation, but is, on the contrary, a part of everyday life experience. It has been calculated that about 30 per cent of all groceries are wasted, thrown away as leftovers or food that has expired. These groceries are just silently thrown into the garbage-can without any accompanying ceremonies or celebratory rituals; it is an all-too-familiar waste of resources and energy which happens every single day, across the globe, concealed by everyday familiarity and an ignorance of the efforts and energy put into those groceries. The contemporary period is one of great and systematic waste and, in the face of environmental disaster, we might need to reflect on these conditions sooner than we think. In other words, we are paying attention here, as Rehn and O'Doherty (2007: 110) advocate, "to less obvious act[s] of expenditure".

The Collège de Sociologie and the critique of utility

In her ethnographic work on Wall Street investment bankers, Karen Ho (2009) points to the elite culture of Wall Street as a key cultural underpinning of the finance sector. The large investment banks primarily recruit from the top elite universities, preferably Harvard and especially Princeton – Yale, the Alma Mater of George W. Bush, is located in the blue-collar city of New Haven and is thus considered to be potentially too liberal, MIT is too "geeky", and Stanford is too far away from lower Manhattan – which provide Wall Street with a steady inflow of "world class intellects". Ho, being a Princeton graduate herself and having work experience of "the street", suggests that Harvard and Princeton graduates are accustomed to elite society and are thus attracted by the sirens' song emanating from Wall Street; finding the new elite community after a few years at Harvard or Princeton is not easily accomplished. However, the recruitment process involves the continuous affirmation of the talent and intellectual capacities of the elite school graduates, and the entire setting strongly emphasizes a lavish elite community. At times, such enacted beliefs and practices boil down to a very concrete choice that needs to be made on the part of the aspiring Wall Street careerists. For instance, notes Ho, when it comes to lunches, the neophytes are not expected to "bring their own lunch" because that does not constitute a "sign of upward mobility" but rather "connotes a lower-class concern with overspending, a relationship with money that was not nonchalant. It sent messages of asocial behavior, as frugality took precedent over going out and buying lunch with colleagues" (Ho, 2009: 120). No matter whether you are busy, stressed out, or simply enjoy having a sandwich in the lunchroom, the Wall Street neophytes are expected to take part in the community of spenders, demonstrating little or no concern for cost or value. They are the true elite of the American capitalist society, on top of the world – why should they spend their lunches chewing on a pastrami and pesto ciabatta at their computer desks? Even though the issue of what to eat for lunch is a seemingly trivial one, anthropological studies of cultures demonstrate that forms of consumption are not an insignificant factor, rather they are representative of the underlying culture (Douglas and Isherwood, 1979; Miller, 1995). The Wall Street neophyte choosing to silently eat a sandwich in front of his/her computer, or worse, socializing with other categories of workers in the office during the lunch hour is not only, in that case, making a gustatory choice but is also, in fact, violating the entire elite culture at the very heart of the Wall Street ideology. Working on Wall Street, one needs to buy into the lifestyle in full – its tedious self-affirmation of the talent and intellectual capacity mobilized, the self-righteous attitude vis-à-vis other industries, the "money is everything" ideology, and the accompanying norms of extravagant consumption – or else one may fall from grace or start to question the rules of the game. Ho thus suggests that excess and waste are written into the ideology of Wall Street investment banking. As Jean Baudrillard (1998a) said of the consumer society, consumption is no longer a choice or a preference during the contemporary era but an *institution*, even a *duty*. The Wall Street neophyte must, then, heed the call of duty and spend significant amounts of money on fancy luncheons and on consuming fashion items that help to signal his/her membership of an elite community.

On the one hand, the contemporary period is obsessed with controlling the efficiency and effectiveness of using input variables. Like no previous historical period, virtually any hour or asset used in organizations is accounted for and return-on-investment ratios are carefully reported and evaluated, making the concept of *performance* a key measure of all things. On the other hand, the demand-side of the equation is less rigorously monitored. As soon as the consumer has purchased the good, the circulation of capital pertaining to that specific item comes to a close, and it is basically up to the consumer to decide how to handle the newly-acquired commodity. The producers of the commodity can only hope the consumer is willing to return in order to buy yet another commodity as soon as possible. These conditions make utility an almost sacred notion on the supply-side of the equation. On the demand-side, other norms and institutions prevail. While production needs to be efficient and financially sound, consumption can take place in any manner: Production is regulated, consumption is free. Once your parents have stopped telling you to eat your vegetables at the end of your formative years, no one tells you how to consume. There are, unquestionably, norms that regulate the domain of consumption, but in comparison with the domain of production, it is basically a land of the free.

The issue of utility was a part of the topics of principal interest to the *Collège de Sociologie* theorists in interwar Paris (discussed in Chapter Two). Being primarily influenced by the works of Marcel Mauss and his forerunner Émile Durkheim, the Collegians were fond of using the dichotomy of the sacred and the profane, as advocated by Durkheim in his thesis on the elementary forms of religion. The sacred is the domain from which a number of practices and myths derive, all sharing the capacity to wield destructive effects on the institution of instrumental rationality, utility in its conventional sense of the term:

> The Collegians envisioned the sacred as a dynamic force propelling collective movements based in feelings that oscillate between repulsion and attraction. Such outbursts can undermine as well as (re)construct the foundations of the social order. For these reasons, the Collège's predilection was to explore social formations – secret societies, religious orders, and political vanguard groups – often bypassed by traditional sociology. (Richman, 2003: 32)

Pearce further explicated the position of the Collegians and Georges Bataille:

> [F]or both Durkheim and Bataille, the sacred invokes feelings of both attraction and repulsion ... but for Bataille it is also linked with instability, with violence and its violent containment; with the cruelty of sacrificing others and with the subsumption of individuals within totalizing group processes when they fearlessly confront death and are willing to sacrifice themselves ... Further, while in contemporary societies sacral processes have become more obscure and suppressed, less obviously religious, they are still present, as can be seen in the way men are attracted to sacrificial ceremonies and festivals. (Pearce, 2003: 3)

The sacred is, thus, the domain wherein the logic of everyday life existence is upended and other rules apply. For Bataille, the domain of the sacred is, thus, most fascinating in terms of providing an escape from the tedious existence of everyday life. In addition, the Collegians were, suggests Richman (2003: 34), principally interested in all sorts of social practices that undermined or violated "utilitarian criteria": "[T]he Collegians were united in their common explorations of myth, power, and the sacred as exemplary areas capable of undermining the hegemony of utilitarian criteria." Bataille's concept of the general economy, advanced as a contrast to what he calls "the restricted economy", is based on such a critique of the norm of utility. Like no other modern social theorist, perhaps, Bataille was interested in forms of transgression such as eroticism, death, violence, and other forms of excess. This makes him a most original, if not peculiar, thinker whose writing is often sketchy, incomplete, and even disturbing. As a social theorist, Bataille remains marginal, but he is still interesting in terms of serving to defamiliarize widely taken-for-granted beliefs and assumptions. "[I]t is human to burn", declares Bataille (1988a: 21), which serves as some kind of slogan for his diverse and largely heterogeneous work.

The restricted and the general economy

The concept of the general economy is based on Bataille's (1988b, 1991) study of historical and anthropological literature on myth, ritual and ceremony in a variety of advanced and "primitive" cultures. As usual, Bataille does not advance his theses very modestly, but instead declares a "Copernican transformation" in his understanding of the economy:

> I will simply state, without waiting further, that the extension of economic growth itself requires the overturning of economic principles – the overturning of the ethics that ground them. Changing from the perspectives of restrictive economy to those of general economy actually accomplishes a Copernican transformation: a reversal of thinking – and of ethics. (Bataille, 1988b: 25)

Previous theories and doctrines of the economy have emphasized the economy on the basis of the principle of scarcity; only through toil and effort will anything be produced that is of any economic value is this world, and the theory of economics is one of blood, sweat, and tears. Bataille rejects such a Puritan view of economic production, instead suggesting that what characterizes various societies is not scarcity but *abundance*, not only careful accumulation but also mindless *waste*: the domain of the economy – the accumulation and distribution of resources at large – is not only concerned with creation but also with *destruction*. In Joseph Schumpeter's (1942) much celebrated work on the dynamic of capitalism, what he calls "the gale of creative destruction" is advanced as the primus motor of capitalist development. Besides this seminal work, relatively few economists and social theorists have been concerned with the apparent destructive side of capitalist accumulation. Bataille summarizes his arguments in a passage worth citing at length:

> We cannot harness any energy without expending some. This is the
> principle of work that is inscribed in a rational society in burning let-
> ters: WHOEVER DOES NOT WORK DOES NOT EAT. But taking hu-
> manity on the whole during a representative average period, humans
> do not need to expend in order to harness the energy that is neces-
> sary to maintain themselves at an already attained level … Humans
> normally expend larger sums of energy than is necessary for them to
> continue. There is regularly an excess which must either be accumu-
> lated – in this case, there is growth (demographic growth, the augmen-
> tation of potential for production) – or *consumed* … It is a fundamen-
> tal but universal error to think that a sum of expendable energy *should*
> be put to some use. On the contrary, there is a sum of energy that *must*
> necessarily have no use. Even the movement of the energy within us,
> taken as a whole, cannot have any other result than *consumption*. If
> you will, wealth, accounting for the utilitarian expenditures without
> which it would not exist, can have no other end than an *unjustifiable*
> expenditure (about which it would be impossible to make any sense).
> (Bataille, 2001: 244)

In Bataille's view, not all energy accumulated can be rationally (i.e., convention-
ally) consumed and thus the so-called accursed share, the part containing the excess energy, needs to be destroyed. This destruction of the accursed share is not,
however, devoid of meaning as it is part of the ritual whereby society and its capacity for accumulating resources are constituted on the basis of such rituals.
Bennington summarizes:

> [A]ny circumscribed system receives more 'energy' from its surround-
> ing milieu than it can profitably use up in simply maintaining its exist-
> ence. Part of the excess (the 'luxury' with respect to what is strictly
> necessary) can be used in the growth of that system, but when the
> growth reaches its limits … then the excess must be lost or destroyed
> or consumed without profit. (Bennington, 1995: 48)

"Bataille's principle of general economy is that there is always excess. It is this
excess that poses the 'economic' problem in that it is what constitutes the wealth
that circulates in this most general economy. The excess has to be put to work
in growth, or it has to be spent in exuberant squandering, or it threatens the very
existence of all beings within the system", writes Lee (2007: 245). Linstead (2002:
670) adds that consumption consists of two parts, a "reducible one", the "subsist-
ence minimum needed for the immediate conservation of life", and the "excess
of energy needed for unproductive expenditure". The latter form of consumption
is not "taken in[to] account by the calculations of the restricted economy" but

operates outside of such utilitarian frameworks. Linstead (2002: 669) contends that "[i]t is *precisely* what we waste, what we produce as excess, that makes us human". McWhorter (2007) suggests that the rules of the general economy are not restricted to material resources but also include cognitive processes such as thinking and the accumulation of shared knowledge. In McWhorter's (2007) view, thinking that deviates from rationalist pathways is, *per se*, a form of squandering:

> Bataille seems to embrace nonreasoning thinking – just undirected, nonproductive mental activity – as part of the exuberant squandering that is the general economy … But even from the point of view of the human species, thinking is rarely necessary; and when it is, it is necessary only in the form of reason – that is, thinking in the service of solving immediate problems. More often than not, mental activities like questioning and wondering are detrimental to the business of survival. Curiosity doesn't just kill cats. In the most rational of possible worlds, ambient thinking would probably be disallowed. (McWhorter, 2007: 158)

The consequence is, McWhorter (2007: 162) argues, that "knowledge belongs, in the end, not to the restricted economy of acquisition and preservation, but to the general economy of luxury and waste". She continues: "In the truest operations, knowledge is not in the service of utility and conservation at all but an event of pure expenditure." Following Bataille's scheme, thinking is either the source of productive growth or exuberant squandering. How, then, to draw the line between the two categories is a matter of outcomes and effects, and ultimately an empirical question.

Bataille's theory of the accursed share may appear frivolous and counter-intuitive but what he seeks to address and render problematic is the idea that human existence is exclusively rooted in utility, in instrumental rationality, in utilitarian reason. "Bataille's target is utility, in its root", argues Baudrillard (1998b: 192). In Derrida's (1998: 123) view, Bataille suggests that there is no *telos* of meaning, no underlying ultimate structure rendering human existence meaningful; on the contrary, it is the participation in the sacred – always in opposition to the profane – that gratifies human existence. Humanity is sustained by "reducible consumption", but such consumption can never be more than that, an indispensable element in the profane world. To constitute humanity, the sacred and its myth, transgression, and the destruction of value need to be attended to. Best and Kellner (1991: 35) summarize Bataille's contribution thus:

> Bataille … championed the realm of heterogeneity, the ecstatic and explosive forces of religious fervour, sexuality, and intoxicated experience that subvert and transgress the instrumental rationality and nor-

malcy of bourgeois culture. Against the rationalist outlook of political economy and philosophy, Bataille sought a transcendence of utilitarian production and needs, while celebrating a 'general economy' of consumption, waste, and expenditure as liberatory. (Best and Kellner, 1991: 35).

It is important here to recognize the situated nature of Bataille's work. The bourgeois society of the interwar period and the emerging fascism were a fertile soil for theorizing the irrational leading up to the devastating outcomes of WWII. In the contemporary, liberal, and essentially secular society is perhaps rendering Bataille's thinking less relevant? As Goux (1998: 198) critically remarks, "productive expenditure now entirely dominates social life. In a desacralized world, where human labor is guided in the short and long term by the imperative of utility, the surplus has lost its meaning of glorious consumption and becomes capital to be reinvested productively, or constantly multiplying surplus-value." Rather than being a real sacrifice in the original sense of the term, as part of the sacred, excess and waste are today – as the example of wasted groceries suggests – more a part of the profane, everyday experience. "No society has 'wasted' as much as contemporary capitalism", remarks Goux (1998: 199). In addition, Bataille's theory – if a theory in the proper sense of the term at all, or merely a set of propositions, a loosely articulated analytical framework – rests on precarious grounds. It is speculative and poorly integrated, but it also offers an interesting idea, i.e. that waste is an elementary part of any economic regime and social system. All societies demonstrate some kind of sacrifice or spending of productive resources – at times barbaric, as in the case of human sacrifice, in other cases more modest, as in the burning of a few candles to celebrate the dead – and waste seems to be, to some extent, bound up with its counterpart, production. It needs to be added that Bataille emphasizes various forms of conscious and self-aware resource destruction, that of carefully planned and orchestrated rituals and ceremonies, while much waste is today bound to be overlooked or ignored, at times subject to moralistic complaints at best, especially if it is subaltern groups or younger human beings who are responsible for the waste. In what follows, the principle of squandering will be used to discuss a few key processes in contemporary society.

The attention economy and the cinematization of society: The overflow of sense impression

The concept of attention

Beller's (2006) cinematic organization of society and its highly heterogeneous forms of visual labor is thus ultimately a matter of attention; consequently, the concept of the *attention economy* (to be discussed shortly) has been introduced as a pertinent label for the most recent economic regime. Attention is a concept closely associated with cognition or even neurology in general. Without reducing the attention economy to a matter of applying learning from cognition science, it is useful to think of cognition not as something that is the outcome of conscious choices and preferences, but as something that in many ways lies beyond the control of individual actors. For instance, in his *Principles of psychology* (first published in 1890), William James (1950: 402) writes "my experience is what I agree to attend to": "Only those items which I *notice* shape my mind – without selective interest, experience is an utter chaos. Interest alone gives accent and emphasis, light and shade, background and foreground, in a word", argues James (1950: 402). However, even if attention is of key importance to cognition, it is *per se* incapable of "creating" an idea, suggests James (1950). Instead, an idea must "already be there before we can attend to it" and attention "only fixes and retains" what is in the consciousness (James, 1950: 450). For James, attention is a mechanism which triggers cognition but which, nevertheless, is not a part of the cognitive structure. Instead, James implies, attention is the nexus between the perceptual apparatus and cognitive structure, inducing various cognitive processes. In this image of cognition, what James famously referred to as a "stream of thought", thinking embedded in attention is a matter of discipline since mind and perception may easily wander and it is not always the case that the perceptual apparatus and the cognitive structure are co-aligned and integrated. As a consequence, both attention and thinking need to be trained and disciplined; seeing is not of necessity under the full control of the individual but gradually becomes so after periods of training. Thus, what Nietzsche spoke of as "the dogma of the immaculate perception" (cited in Bourdieu, 1993: 219) also holds true for James; perception is never separated from memory and thinking, and our capacity to perceive and observe new things is bound up with what we have previously seen. Crary (1999) explicates James' position:

> Working with a more act-oriented term 'thought' instead of 'consciousness', James uses the image of the stream to describe the fundamentally transitive nature of subjective experience – a perpetually changing but continuous flow of images, sensations, thought fragments, bodily awareness, memories, desires – which he sets against the older or even contemporary accounts for which consciousness has discrete contents and elements. (Crary, 1999: 60)

As more contemporary students of perception and cognition have shown, "vision … is a dynamic process in which the brain, largely automatically, filters, discards, selects and compares information to an individually stored record" (Stafford, 2009: 277). As a consequence, assuming a dynamic and recursive relationship between perception and cognition, "[c]onsciousness apparently produces its own content, i.e., the world," contends Stafford (2009: 281). Perception and vision more specifically are thus rather precarious sources for knowing the world. The eye deceives and previous experiences in many ways lie in-between ourselves and the world we are trying to understand. Prejudice, group thinking, past experiences, perceived inconsistencies, and cognitive dissonance are some of the factors to consider when it comes to making adequate and informed judgments within a specific field. As has been demonstrated by organization theorists, what Herbert Simon (1957) named *bounded rationality*, an umbrella term accommodating such mediating factors, strongly influences managerial behavior and action:

> [T]op managers, who are bolstered by memories of past successes, usually cling to their cognitive structures and may misperceive events and rationalize their organizations' failures … Such rigidity often leads to crises, accompanied by superficial remedies and delays. Instead of examining the deeply held values, executives may initiate minor revisions to certain operational issues. As crises intensify, rigidity of beliefs intensifies, limiting the firm's response to the forces of change causing the organization to eventually collapse. (Tsang and Zahra, 2008: 1448)

In decision-making, past experiences and relatively simple causal relations play a key role:

> Organizations interpret present states as outcomes of past decision. They base decisions on causality assumptions … Since an organization is unable to operate on the basis of knowledge containing many contingencies it has to keep its picture of the world – its assumptions concerning means and ends – simple (which might explain managers' aversion to scientific knowledge pointing to many contingencies and their preference for consulting knowledge). (Kieser and Leiner, 2009: 522)

Here, Maurer and Ebers (2007: 275) use the term "cognitive lock-in" to denote cases where managers are unable to think outside the proverbial box or, even worse, remain unaware of their biases and preferences. The work of Karl Weick – in many ways strongly indebted to William James – has been very instructive in pointing out that managers and other organization members (e.g., the smoke-jumpers examined by Weick, 1996) are trained and disciplined to think of their work and themselves in specific terms and professional ideologies; thus, when situations call for a reevaluation of such terms and ideologies, it is conspicuously difficult for them to execute what Argyris and Schön call *double-loop learning*, the "[o]rganizational inquiry which resolves incompatible organizational norms by setting new priorities and weightings of norms, or by restructuring the norms themselves together with associated strategies and assumptions" (Argyris and Schön, 1978: 24, original emphasis omitted).

Cinematization

In contemporary society, the abundance of data, information, messages, and images far exceeds the cognitive capacities of the members. Manovich (2009: 320) says that, on Facebook, 14,000,000 image uploads are made daily and that 65,000 new videos are uploaded to YouTube every twenty-four hours. The concept of information overflow is justified. In addition to the mass of information produced, stored, and retrieved on an everyday basis in society, much of that information is increasingly circulating in the form of images and film clips. There are thus two major shifts in how we consume and make use of media is today's society: First, we use digital media, enabling endless possibilities of manipulating and distributing documents and files. Second, to a larger extent, everyday communication is based on what Vilém Flusser (2002) calls *surface media*, images and cinematic representations, operating along different syntactic and cognitive structures than linear media (e.g., written text). This "cinematization of society" and its reliance on very flexible and malleable digital media will be discussed as key elements of what has been called "the attention economy". This attention economy is primarily based on the principles of abundance and squandering, the continuous and ongoing production and consumption of images.

The German media theorist Friedrich Kittler (1990: 116) examines the nineteenth century when the surfaced media we are immersed in today were first developed. When contrasting the years 1800 and 1900, the advancements of the century become conspicuous:

> Besides from mechanical automations and toys, there was nothing [the year 1800]. The discourse network of 1800 functioned without photographs, gramophones, or cinematographs. Only books could provide serial storage of serial data. They had been reproducible since Gutenberg, but they became material for understanding and fantasy only when alphabetisation had become ingrained. (Kittler, 1990: 116)

During the nineteenth century, a wide range of new media were developed, in many cases based on electronic engineering. In combination with the emergence of a new and previously unseen visual culture, the nineteenth century developed the modern, using Kittler's (1990) term, *discourse network*: "The term discourse network ... designate[s] the network of technologies and institutions that allow[s] a given culture to select, store, and process relevant data. Technologies like that of book printing and the institutions coupled to it, such as literature and the university, thus constituted a historically very powerful formation, which in the Europe of the age of Goethe became the condition of possibility for literary criticism". (Kittler, 1990: 369). Kittler suggests that media (e.g., the printed book, radio, cinema, the digital computer) strongly influence society in term of providing new possibilities for storing and retrieving data and information, but also in terms of enabling and legitimizing new ways of thinking. The shift from the scroll to the printed book shifted the focus from the scholastic tradition of knowing, during the medieval period, i.e. to only study and comment on a few canonical texts (Aristotle, the Church Fathers), to developing a broader and more innovative relationship with the texts investigated. With the printed book, the modern view of the author is born, as are accompanying concerns regarding originality and plagiarism (Eisenstein, 1983; Johns, 1998; Febvre and Martin, 1997). It has also been suggested that the very Cartesian image of the self as an autonomous and self-reflective cognitive universe, capable of thinking and critically reflecting on the state of things, is bound up with the very practice of reading; only when spending time in isolation to decode and understand linear texts can the subject be constituted as an autonomous being (Bolter, 1996) – this peculiar and highly idiosyncratic Western conceptualization of the self. In the more recent shift to digital media, beginning in the 1980s, but very much taking off in the 1990s with the more widespread use of the Internet, increasingly providing information in a cinematic form, linear media no longer serve as the single most important storage of collective memory; instead, various forms of images dominate.

Criticizing new media is far from a recent phenomenon. As Sconce (2000) points out, virtually any new media have been subject to worries and concerns, beginning with Plato's rejection of writing in *Phaedrus*, when suggesting that, in comparison to speech, writing is little more than a supplement that is even detrimental to human memory and imagination. At the same time, the more sophisticated critiquing of media is an inherently modern phenomenon. The nineteenth century was a period of swift migration (e.g., from the countryside into the towns and cities, from Europe to North America) and urbanization; theorists such as George Simmel, Sigfried Kracauer, and Walter Benjamin, theorists of the early twentieth century and of the Weimar Republic in Germany, were all interested in the new social changes. Simmel, a Berliner, living in a city of 420,000 inhabitants in 1850 and 2 million in 1900 (Gay, 1984: 49–50), was interested in how this exposure to massive amounts of impressions is to be coped with. Simmel (1971: 326) thus talked about "the metropolitan type" who is able to avoid responding emotionally to all kinds of sense impressions; instead "[t]hose events are moved to a sphere of mental activity which is least sensitive and which is furthest removed from the depths of the personality". Simmel was addressing here the lack of emotional responses on the part of the urban subject as a "blasé attitude" whereby "[t]he nerves reveal their final possibility of adjusting themselves to the content and the form of metropolitan life by renouncing the response to them" (Simmel, 1971: 330). The emotional detachment, or even indifference, that has been associated with the *Gesellschaft* of urban life forms – the tough-minded and hard-to-impress attitudes that the newcomer to the city rarely fails to notice, the essence of *urban cool* – is explored and theorized by Simmel.

In addition to such demands to cognitively and emotionally separate oneself from the "buzzing, blooming confusion" (using William James's phrase) of city life, Siegfried Kracauer (1995) examined the emerging visual culture of the urban setting. While rural life has primarily been concerned with work and a few leisure activities, leaving intellectual pursuits to the Church and making texts and images relatively marginal phenomena in human lives, the modern city is overflowing with images, messages, and texts that need to be responded to in one way or another. The use of billboards and other forms of visual communication in the major metropolitan areas was, thus, a relatively new phenomenon. Moreover, the photograph, another medium developed in the nineteenth century and used to document, for instance, the American civil war in 1861–1865, was widespread during the early twentieth century. For Kracauer (1995a: 50), the widespread use of photography is not an inherently positive tendency because the photography is, in contrast to memory being "full of gaps" and inconsistencies, capable of capturing "the entire spatial appearance of a state of affairs"; the photography excludes nothing (Barthes, 1981). For Kracauer, memory images retain that which has *significance*, that which matters, that which is important. Photographs, on the other hand, provide the full temporal and spatial presence of a situation, giving the impression of providing an "objective" and self-explanatory view of a particular situation; however, photographs are never capable of explaining anything inasmuch as they do not "encompass the meaning to which they refer" (Kracauer, 1995a: 50–51). Photographs make memory appear to be fragmented and incomplete since they omit nothing from the scene; however, this perceived completeness of the photograph is deceiving, suggests Kracauer, because images need to be accompanied by a broader understanding of what they portray. As Kracauer (1995a: 53) and many others have remarked, "until well into the second half of the nineteenth century, the practice of photography was often in the hands of former painters", testifying to the photograph in terms of not being something that mirrors reality, but a new medium for comprehending reality. However, when the meaning of the photograph gradually slides from being an artistic representation to a form of "mediated truth", the photograph needs to be critically evaluated. In this new situation, the new medium serves to exclude more critical reasoning and understanding. Expressions like "see for yourself!" or "a picture is worth a thousand words" suggest that visual perception is a reliable source of undisputed information, that visual inspection is an epistemologically-solid ground for knowing. Kracauer (1995a) strongly rejects such beliefs and ideologies, saying instead that the images themselves need to be explained since, in many cases, they conceal rather than reveal situations and events. Images are spurious in terms of giving the *impression* that they are epistemologically uncomplicated, while actually being embedded in a variety of forms of knowledge in order to fully make sense.

More than half a century after Kracauer, Susan Sontag (1973) expressed similar concerns regarding the use of images and photographs in, for instance, propaganda and news reporting; photographs are, argues Sontag (1973: 23), "inexhaustible invitations to deduction, speculation, and fantasy". The consequences of the wide-spread use of photographs and other visual media are significant; Sontag (1973: 24) says: "Industrial societies turn their citizens into image-junkies; it is the most irresistible form of mental pollution." Largely echoing Kracauer's concern regarding the popularity of *Bild Zeitungen* and *Kino* (cinema), as well as other cultures associated with Americanism in Germany during the Weimar era (Hansen, 1995: 367), Sontag is conspicuously explicit regarding her critique of the role of the image. For both Kracauer and Sontag, in the situation where the image is a substitute for the text, critical reflection is in decline:

> Never before has a period known so little about itself. In the hands of the ruling society, the invention of illustrated magazines is one of the most powerful means of organizing a strike against understanding. Even the colorful arrangement of the images provides a not insignificant means for successfully implementing such a strike. The *continuity* of these images systematically excludes their contextual framework available to consciousness. The 'image-idea' drives away the idea. The blizzard of photographs betrays an indifference towards what the things mean. (Kracauer, 1995a: 58)

"Stupid and unreal films are the *daydreams of society*, in which its actual reality comes to the fore and its otherwise repressed wishes take on form", concludes Kracauer (1995c: 292), in many ways anticipating the Frankfurt School criticism of the "culture industry" (Adorno, 1991), the industrialization of mass culture.

With the foundational work of Simmel, Kracauer and others, the view of modernity shifts from being one of engineered spaces of concrete and steel, of skyscrapers and railways (Bauman, 2005: 62), to what Singer (1995: 72–73) called "a *neurological* conception of modernity". Modernity is, then, both the construction of new life worlds strongly shaped by the newly-developed technologies and buildings that enable new forms of transportation and dwelling and new cognitive structures. Moderns do not only need to be able to handle train journeys and elevator rides, but also to cognitively and emotionally sort out and structure large amounts of visual data and information. Singer (1995) explains:

> These theorists [Simmel, Kracauer, *et al.*] focused on what might be called a *neurological* conception of modernity. They insisted that modernity must also be understood in terms of a fundamentally different register of subjective experience, characterized by the physical and perceptual shocks of the modern urban environment … Modernity implied a phenomenal world … [that] was markedly quicker, more chaotic, fragmented and disorienting than in previous phases of human culture.

Of all the media to have played very important roles in instituting this new neurological regime, cinema – the "motion picture" – is undoubtedly the most important innovation. During the nineteenth century, a new visual culture was developed, fuelled by the department stores, world expositions, melodrama, phantasmagoria, wax museums, morgues, and so forth that were being developed in order to both exploit and satisfy the *scopophilia* of the era. In Hansen's (1995: 363) view, cinema is to be seen as the culmination of two parallel trajectories in the nineteenth century, i.e. those of consumerism and visual culture, wedded together in the new medium revolutionized by the French Lumière brothers. In cinema, "sensations, fashions, and styles" (Hansen, 1995: 363) are brought into a new form of visuality, the cinematization of perception. Film theorists such as Beller (2006) have emphasized that this cinematization is an ongoing process whose full consequences we are still unable to oversee, but for Beller, the shift from linear media to cinematic media is a major historical event:

> Cinema is the development of a new medium for the production and circulation of value, as important in the reorganization of production and consciousness as the railroad track and the highway. Human endeavours generally grouped together under the category 'humanities', and (perhaps) once experienced as realms of relative freedom, can be and are being figured as economically productive. The entire history of cinema remains as a testament to the practice; advertising, television, and culture generally *testify* to it. (Beller, 2006: 207)

Using a Marxist framework of analysis and terms such as "the cinematic organization of society", "the production value of attention", and "visual labor", Beller advances visual practices as an important factor in the production of value in the contemporary capitalist regime. "[T]he cinematic organization of society has led to the restructuring of perception, consciousness and therefore of production. Furthermore, this production vis-à-vis the cinematization of society has become increasingly dematerialized, increasingly sensual and abstract, to the point it occupies the activities of perception and thought itself", says Beller (2006: 295). All the economic activities derived from the Internet (online trade and shopping, computer games, the dating industry, etc.) are in various ways based on the visual labor of the community; economic value is acquired through the attention of the general public and even in cases where services are given away for free (an important element of, for example, Google's business model), economic value is generated elsewhere in terms of having know-how regarding what people pay attention to.

Attention and economic value

We learn from William James and others that attention is what drives understanding and meaning, but attention is increasingly becoming a scarce resource in a society overflowing with visual media, and sense impressions more broadly. In the contemporary economy, attention is thus, perhaps, the single most important capacity to exploit. We are, some writers propose, living in the *attention economy*. The attention economy is an evocative term. Lanham (2006) contrasts the terms *information economy* and *attention economy* and suggests that we are moving into a situation where it is the capacity to attract attention that is the new production factor *par excellence*: "[I]nformation is not in short supply in the new information economy. We're drowning in it. What we lack is the human attention needed to make sense of it all. It will be easier to find our place in the new regime if we think of it as an economist of attention. Attention is the commodity in short supply" (Lanham, 2006: xi) Eriksen (2001: 21), too, stresses attention as a "scarce resource" in contemporary society: In "the information society, the scarcest resource for people on the supply side of the economy is neither iron ore nor sacks of grain, but *the attention of others*. Everyone who works in the information field – from weather broadcasters to professors – competes over the same seconds, minutes, and hours of other people's lives." Google, perhaps the most widely discussed business phenomenon in terms of growth, accumulation of wealth, and cultural significance over the last few years, is one example of a company which thrives on the attention economy and which makes money on the basis of its capacity to monitor and detect what people pay attention to:

> For Google, being in the attention business means it's of utmost importance for Google to find out what people are paying attention to. To do that, Google has made a history of giving away services that other companies charge an arm and a leg for. Even more amazingly, the services that Google gives away is usually better than the services that other people are charging for … In exchange for that free services and soft-

ware, Google wants to look over your shoulder and take note on what you are paying attention to and what you are ignoring. (Mitch Wagner, Google, cited in Beller, 2006: 303)

The line of demarcation between being attended to and being ignored may make the difference here between success and failure. As a consequence, companies pay companies like Google to be noticed when browsing the Internet and using search engines to navigate it. Lanham (2006: xi–xii) emphasizes here that the attention economy is a particular capitalist economic regime wherein "stylistic devices" are playing an increasingly larger role: "The devices that regulate attention are stylistic devices. Attracting attention is what style is about. If attention is now at the center of the economy rather than stuff, then so is style. It moves from the periphery to the center. Style and substance trade places." "Style displaces substance", suggests Lanham (2006) rather provocatively. Rather than making such declarative statements in order to announce the shift in history, one might recognize that Lanham is rehabilitating the field of aesthetics as a source of know-how and competitive advantage in contemporary capitalism.

Returning to Bataille's concept of the general economy, the very idea of the attention economy is that there is a surplus, an excess of information (experienced as a "shortage of attention") during the contemporary period. Even though more than eight billion people live on this planet, it is still a challenge to attract the attention of any one of them and thus the professional field of marketing – practically, the "science of commercial attention" – is thriving as a large number of firms, companies, organization, agencies, etc. are seeking to attract the attention of a general public burdened by the constant noise of the various messages and information distributed and piped into virtually every domain of everyday life. While Bataille suggested that in historical societies the accursed share represented only a relatively minor part of the resources accumulated, in the contemporary society, that of the attention economy of competitive capitalism, a significant number of resources are wasted, squandered, and destroyed on an everyday basis. One single Sunday issue of the New York Times contains more factual information than Elizabethan England's largest library (Davenport and Beck, 2001: 4), and there is a global and continuous production of news, information, and data that will never be interpreted or used in any meaningful or productive way. In the restricted attention economy, all information would find its recipient who would be capable of understanding it and draw conclusions on the basis of it; however, in the general attention economy – or the attention economy in short, to avoid any unnecessary truisms because the attention economy is by definition part of what Bataille calls the general economy – there is a constant overproduction of information aimed at human beings that is impossible to sort out and structure, let alone fully understand and oversee. For instance, news agencies, as Czarniawska (2009) suggests, not only filter the inflow of news but also actively construct it. Rather than investigating the accuracy of incoming data and information, in an attempt to make inferential analyses on the basis of the available information, it is more convenient for journalists to expand and report on some of the more easily-understood events. As, for instance, Niklas Luhmann (2000) has suggested, the mass media is a self-referential autopoetic social system producing its own communication more or less in detachment from other social systems. Paul Virilio has also remarked, in countless publications, that, in the high-speed society, there is no longer any time for reflection (e.g., critical evaluation and analysis) and, consequently, there is little left to share than emotions:

> Today, we have no longer time to reflect, the things that we see have already happended … Is a real-time democracy possible? An authoritarian politics, yes. But what defines democracy is the sharing of power. When there is not time to share, what will be shared? Emotions. (Virilio, 2002: 43)

It is then hardly surprising that the tabloids and, increasingly also, the more highbrow publications are being filled with alarmist announcements and emotionally-charged stories, about individuals who are either fortunate or unfortunate, which demand little or no journalistic expertise to publish. The attention economy is certainly a high-speed economy embedded in what Beller (2006: 295) speaks of as "the historico-material shattering of the linguistic alongside the rise of the visual" and "the reconfiguration of the imaginary and the subject in the logic of the spectacle". In Derrida and Stiegler's view (2002: 103–104), the contemporary period is characterized by a return to the hieroglyphic, ideographic, and pictorial modes of communicating; television, video, and the cinema have reintroduced the pictogram as the predominant element of communication. In their continuous exposure to images and film clips, individuals need to shut down their perceptual and cognitive capacities in order to avoid losing their focus on what truly matters; thus, there is an unceasing and ongoing squandering of resources in the images and film clips, and other forms of information (wherein written texts naturally still play a dominant role). During the contemporary period, economic agency rests on a high tolerance for this kind of waste since it is instituted within the contemporary economic regime of accumulation.

Attention, style, and economic value: The case of fashion

In the attention economy, style is a distinguishing feature that makes individuals who are increasingly overburdened by visual messages pay attention. The concept of style is closely associated with aesthetics and fashion, two terms brought to the forefront of contemporary capitalism. While neoclassic economic theory has primarily been concerned with the production of commodities and services, during the present period, the focus has, if not shifted, at least been complemented by theories of consumption. The social critic and activist Guy Debord, leader of the Situationist Movement and author of the minor classic *Society of the spectacle*, points to this shift in emphasis in bourgeoisie culture:

> For classical capitalism, wasted time was time that was not devoted to production, accumulation, saving. The secular morality taught in bourgeoisie schools has instilled this rule of life. But it so happens that by an unexpected turn of events modern capitalism needs to increase consumption, to raise 'standards of living' ... Since at the time production conditions, compartmentalized and clocked at the extreme, have become indefensible, the new morality already being conveyed in advertising, propaganda and all forms of the dominant spectacle now frankly admit that wasted time is the time spent at work, which later is only justified by the hierarchized scale of earnings that enable one to buy rest, consumption and entertainment – a daily passivity manufactured and controlled by capitalism. (Debord, 1981: 73)

In contemporary capitalism, individuals may not be needed as producers since unemployment rates are stable at around 10 per cent in many OECD countries, but individuals are still needed as consumers. One of the mechanisms upholding consumption is the concept of fashion, the temporal ebbs and flows of certain ideas, styles, and arrangements of garments, technologies, and so forth. Fashion has always been a part of human societies but its influence and economic importance have been further accentuated during late modern competitive capitalism. As a concept of social theory, fashion is perhaps best associated with the work of George Simmel and with Thorstein Veblen's (1994) *Theory of the leisure class* and what he calls *conspicuous consumption*, the strategic signaling of access to economic capital through, for instance, the wearing of lavish dresses and expensive jewelry which demonstrate the social and economic status of a particular class. Complementing Veblen's analysis, Lipovetsky ([1994] 2007) suggests that modern fashion is in line with what Max Weber speaks of as the rationalization and de-traditionalization of society inasmuch as female dresses were, from the 1920s, increasingly being designed on the basis of their utility – the in many ways revolutionary designs of Coco Chanel being one example used by Lipovetsky – rather than their symbolic functions:

> Through the cult of sports, the prototype of the slim, svelte, modern woman who plays tennis and golf took over from the sedentary woman hobbled by her flounces and lace. The simplification of clothing in the

1920s, the elimination of gathers and frills in favour of restrained, clean lines, was a response to this new idea of lightness and energy drawn from sports. (Lipovetsky, [1994], 2007: 82)

In a trajectory in many ways comparable with modernist architecture revolutionized by a generation of architects rejecting ornamentation – Adolph Loos spoke of ornamentation as "the sadism of the eighteenth century" (cited in Klein, 2004: 10–11) – the female dress served less to signal wealth and social standing but rather served to establish the "modern woman" as an active or even enterprising agent (see, for instance, Entwistle, [2000], 2007).[25]

While much scholarly work is skeptical, or even condemns fashion as being frivolous, wasteful, and generally problematic, it is also possible to take a somewhat more moderate view, taking into account fashion's ability to effectively combine renewal and stability, playfulness and predictability, imitation and innovation. Individuals tend to live fairly repetitive lives ordered along a few recurrent events and occurrences, with life being cyclically structured around seasonal variations of the year, collective ceremonies, and family celebrations (birthdays, wedding days, graduation days, etc.). At the same time, many human beings appreciate novelty and creativity, and fashion enables the combination of a fairly repetitive and predictable life with opportunities for the creative renewal of the self (see, for instance, ten Bos, 2000: 11). In one of the earliest accounts of the role of fashion in social theory, Edward Sapir ([1931], 2007) points to how fashion balances the need for conformism and individualism in modern society:

> Fashion is custom in the guise of departure from custom … Fashion is the discrete solution of the subtle conflict. The slight changes from the established dress or other forms of behaving seem for the moment to give the victory to the individual, while the fact that one's fellows revolt in the same direction gives one a feeling of adventurous safety. (Sapir, [1931], 2007: 40)

Contrary to such a view, conceiving of fashion as liberating and innovative, Blumer ([1969] 2007: 233) emphasizes that fashion, once it operates, "assumes an imperative position": "It sets sanctions of what is to be done, it is conspicuously indifferent to criticism, it demands adherence, and it by-passes as oddities and misfits those who fail to abide by it." Lipovetsky ([1994] 2007: 82), too, em-

---

25 Film theorist Laura Mulvey remarks here that "fashion now is not what is fashionable, what is fashionable is the actual sculpting of the body itself" (Laura Mulvey, cited in Sassatelli, 2011: 134). The capacity to maintain bodily control and to sculpt the body through hard physical training is what matters and fashion is merely, in Derrida's (1987) Kantian vocabulary, that which serves as the *parergon*, the frame, of the *ergon*, the actual work – the disciplined material substratum, the muscular and fit body testifying to the subject's capacity to execute bodily control.

phasizes this dual nature of fashion: "[F]ashion simultaneously freed personal appearance from traditional norms and imposed on all and sundry the ethos of change, the cult of modernity. Fashion was more than a right; it had become a social imperative." For sociologists, stresses Blumer ([1969], 2007), fashion is thus of key importance as it guides and regulates social behavior. In addition, Blumer predicts that the role of fashion will increase in the future: "Fashion should be recognized as a central mechanism in forming social order in a modern type of world, a mechanism whose operation will increase" (Blumer, [1969], 2007: 245). Ignoring or trivializing fashion as a social mechanism is overlooking its normative and regulatory role in everyday life:

> [T]he fact is clear that fashion is an outstanding mark of moderns civilization and that its domain is expanding rather than diminishing. As areas of life come to be caught in the vortex of movement and as proposed innovations multiply in them, a process of collective choice in the nature of fashion is naturally and inevitably brought into play. (Blumer, [1969], 2007: 243)

In Blumer's account, fashion is not a peripheral, or inherently liberating mechanism, but serves to institute a form of voluntary conformism reconciling structure/agency problems that need to be studied and understood by sociologists.

Siegfried Kracauer, drawing on the work of George Simmel, spoke of the middle-class as the principal drivers of fashion, the social group in greatest need of both demonstrating social standing and introducing new conventions both to respond to and to demonstrate an awareness of:

> Which group within society will become the primary supporter of fashion? The middle classes. The lower classes are immobilized by the weight of their economic burden; the highest classes, by their conservative attitude. The need to stand out grows in proportion to the density of cohabitation; that is why fashion is a metropolitan phenomenon. (Kracauer, 1995c: 247)

Max Weber (1962) defines fashion in terms of resting on novelty rather than "convention", underlining the bourgeoisie demand for new means of expressing social standing and entrenched economic positions:

> Fashion as distinguished from usage exists where the conduct in question is motivated by its novelty rather than by its long standing, as is true with custom. Fashion belongs to the neighbourhood of 'convention' since it originates for the most part from the desire for social prestige. (Weber, 1962: 67)

In the view of both Kracauer (1995) and Weber (1962), fashion accommodates a specific temporality, the careful separation of "now" from "then" which the subject needs to be aware of. Agamben (2009: 47) suggests that fashion needs to be defined on the basis of the temporality that it introduces into human and social life, an institutionalization of certain norms and standards for when a fashion begins and ends: "Fashion can be defined as the introduction into time of a peculiar discontinuity that divides it according to its relevance or irrelevance, its being-in-fashion or no-longer-in-fashion." Fashion is not, then, external to social organization but becomes something that is capable of structuring collective temporal experience inasmuch as certain fashions come and go over time; time is inscribed into the clothing worn in the public sphere. Over the course of history, fashion and shopping more generally have been highly gendered, associated with femininity and passive consumption, a form of frivolous and playful (female) waste of accumulated resources derived from the (male) domain of production. Rappaport (1995), discussing the opening of Gordon Selfridge's new department store on 15 March 1909 in London's West End, stresses these ephemeral and feminine connotations of shopping:

> [S]hopping has long been associated with women, but the meaning of this activity was by no means stable. For much of the Victorian era, shopping had been often denigrated as a wasteful, indulgent, immoral, and possibly disorderly female pleasure. Selfridge and other Edwardian entrepreneurs rewrote the meaning of this indulgence. They used publicity, particularly the print media, to turn disorder and immorality into legitimate pleasures. Shopping was advanced as pleasurable and respectable precisely because of its public setting, which Edwardian business presented as a context for female self-fulfillment and independence. Selfridge addressed women not only as urban actors but also as bodies to be satisfied, indulged, excited, and repaired. Shopping, he repeatedly stated, promised women access to a sensual and social metropolitan culture. (Rappaport, 1995: 131)

For instance, Emile Zola, being attuned to his times, published the novel *The Ladies' paradise* (1881) in order to explore the new culture emerging in the great Parisian department stores such as *Le Bon Marché* and *Samarithain*, which opened in the first half of the nineteenth century (see Miller, 1981). In Zola's view, shopping is an immersive experience involving all the senses and strongly associated with the good life of the urban middle and upper classes.

Taken together, fashion and its accompanying social practice of shopping – originally on the fringes of the economy, but gradually moving toward its center – are in many ways associated with the frivolous and unproductive spending of economic resources. Perhaps it is the very idea of fashion in terms of being, in ten Bos' (2000: xiii) words, "volatile, plural, aesthetic, stylish, superficial, real, experimenting, emotional, embellishing, playful, loving, uncontrollable, ugly, disenchanting, creative and so on", that makes it appealing to humans. In many ways, fashion is connected with the accursed share, the part of the accumulated economic resources that needs to be squandered, the surplus, the excess to be put to use. At least, any activity associated with fashion and shopping rests on the notion of what is glamorous, extravagant and, ultimately, *extraordinary*. The pirated Louis Vuitton bags being sold by street vendors in European cities may not be associated with glamour, but still this specific low-end trade seeks to squeeze the last drop of prestige out of the iconic "LV" logo reproduced unlawfully on these items; it is a conspicuous example of the trickle-down theory of style in the contemporary attention economy, a gradual migration and depreciation of the value of a logo down to the level of mass-produced pirate copies.

Organizing fashion work

Entwistle and Racamora (2006) report on a study of London Fashion Week, an annual event when the international, and primarily British, fashion industry meets and presents its recent work. Just as in the fashion industry more broadly, London Fashion Week is a stage where what Entwistle and Racamora (2006) refer to, with reference to Pierre Bourdieu, as "fashion capital" is both consumed and reproduced. Fashion capital is the accumulated and perceived credibility, as possessed by any actor in the fashion field, ranging from star designers and fashion magazine editors (the real authorities in the field) to upcoming designers or designers in decline, to fashion industry workers of the lower ranks, to "expert consumers". Fashion capital is that which is both spent and acquired during London Fashion Week, and there are specific and highly encoded rules regarding how this capital is circulated. For instance, when seen at a specific show, preferably that of a star designer, an individual will acquire capital. On the other hand, minor designers or fashion houses may receive significant attention (or at least, their fair share of gossip) if an undisputed authority, for example the editor of one of the major fashion magazines, were to attend the show and thus spend some (in most cases) of her time and fashion capital on this show because she expected to see something qualitative and interesting. If the editor of Vogue were to attend the show of a new and upcoming designer, then he or she would instantly acquire additional fashion capital. During London Fashion Week, and in the fashion industry more broadly, there is a most intricate social algorithm used to determine how fashion capital is distributed. Among the many things determining access to fashion capital is the bodily performance of the actors, suggest Entwistle and Racamora (2006: 748): "[I]n the field of fashion, it is critical that one's body articulates fashion capital, position and status in the field." London Fashion Week and other similar events, such as their counterparts in Paris or New York, are carefully designed and orchestrated events that are associated with creativity, glamour, extravaganza, high society and celebrity culture, and innovation. The brute economic interest is kept in the background as though such conspicuous economic concerns would threaten to shatter the image of the fashion industry as a socially-privileged domain celebrating the beauty of life. By and large, London Fashion Week is a social arena constructed on the basis of, and bound up with, the very idea of conspicuous consumption and waste as key elements of contemporary culture.

The fashion industry never claims to be "important" (despite the employment it generates) in the same manner as the energy sector, or the same way the pharmaceutical industry seeks to justify itself on the basis of its social contributions. On the contrary, the fashion industry is in no need of justification as it operates on the register of desire; the desire to take part in a form of consumption that promises little more than the joy and playfulness in temporarily upending the repetitive nature of everyday existence. The fashion industry is, to some extent, the late modern version of the festival, the intentional and strategic consumption of economic resources that extends beyond the domain of mere utility. Fashion is not exactly *needed* and, consequently, it has consistently been condemned or criticized for being irresponsible, immoral, or even irrational; nevertheless, it is highly attractive to humans inasmuch as it imposes, as Agamben (2009) says, a temporal structure on our existence.

In summary, then, fashion is not to be treated as a marginal phenomenon but as something at the very heart of relations in the attention economy; its principal feature is style and its function is to regulate social relations while introducing a sense of playfulness and change into what is otherwise repetitive and recurrent. Underlying these functions is a critique of utility in the tradition of the *Collège de Sociologie* theorists, emphasizing human life not only as the ceaseless advancement of useful and practical resources but also as the squandering of resources, the mindful destruction of what has been accumulated. Just as nature overflows with energy that seemingly goes to waste – e.g., junk DNA, potentially inactive and inert – so too does human existence rely on the excess of resources never used productively. Unfortunately, the present predominant economic theory is primarily concerned with the supply-side rather than with the demand-side of the economic equation, generally being poorly equipped to comprehend anything but scarcity and diminishing returns.

SQUANDERING IN ORGANIZATIONS

Information collection in decision-making

The principle of squandering may appear to be an unexplained element of irrationality in an otherwise rational regime of thinking. As Michel Serres (1995) remarks, humans are accustomed to thinking of their life world as essentially structured and well-ordered and where noise only subsist on the fringes; not so, insists Serres – being is chaotic and noise prevails and only due to the tedious work of humans are there small pockets of order and stability. Work and management lead to the establishment of such domains of control in an otherwise chaotic existence. Similarly, the overarching belief is that modern life effectively uses resources and that there is little or no room for wasteful expenditure. Such beliefs or ideologies (e.g., the neoliberal doctrine that markets never waste economically valuable resources. Peet, 2007: 79) are "infrastructural" inasmuch as the ongoing and continuous waste of resources is overlooked *qua* waste in organizations and in society more broadly; waste is a necessary expenditure to "make the system work". Unemployment (the waste of human capital) is necessary for the overall effectiveness of the economic system – Marxists would say that a necessary level of unemployment keeps down expectations regarding pay – and the squandering of energy in, for instance, transportation due to car driving is necessary for the reduction of the time dedicated to transportation. Squandering, expenditure, and waste are concealed by commonly-shared beliefs and assumptions. In Bataille's general economy, squandering is never inherently wasteful, but has a key role in consuming what is surplus to the economic and social system and what cannot be channeled into productive investment.

The challenge for the organization theorists, then, is to be able to study expenditure in organizations without succumbing to moralistic beliefs regarding the use of scarce resources or pointing to everyday work life expenditure and its productive functions. For instance, Martha Feldman and James March's (1981) study of the collection of excess information in decision-making procedures is exemplary in both recognizing the waste of time and energy while also pointing to the functional role of such seemingly irrational procedures. Feldman and March (1981: 174) summarize their findings in six points which are worth citing at length:

> (1) Much of the information that is gathered and communicated by individuals and organizations has little decision relevance. (2) Much of the information that is used to justify a decision is collected and interpreted after the decision is made, or substantially made. (3) Much of the information gathered in response to requests for information is not considered in the making of the decisions for which it was re-

quested. (4) Regardless of the information available at the time a decision is first considered, more information is requested. (5) Complaints that an organization does not have enough information to make a decision occur while available information is ignored. (6) The relevance of the information provided in the decision-making process to the decision being made is less conspicuous than is the insistence on information, in short, most organizations and individuals often collect information that they use or can reasonably expect to use in the making of decisions. At the same time, they appear to be constantly needing or requesting more information, or complaining about inadequacies in information. (Feldman and March, 1981: 174)

In the behavioral theory of decision-making advanced by Herbert Simon and his collaborators, including James March, decision-making is inextricably bound up with the social, cultural, and cognitive conditions under which the decision is being made. Since most decisions are conducted within the horizon of uncertainty and ambiguity, there is a proclivity to collect more information than is actually used when eventually making them. Beliefs, norms, "gut feelings", and similar non-informational sources are drawn on and, in many cases, decisions are outcomes of debates and controversies rather than the analysis of factual conditions and the perceived need for action. In order to navigate in these domains, there is a need to balance what Brunsson (1982) calls *action rationality* and *decision rationality*, two alternative regimes of decision-making whereby one (action rationality) favors outcomes and tangible effects while the other (decision rationality) strongly emphasizes the importance of adhering to prescribed decision-making procedures. According to Brunsson, the balancing of the two rationalities – a form of merging the two into a situational rationality, one might say – demands a certain form of self-deception, or even hypocrisy on the part of the members of the decision-making community (Brunsson, 1985, 2006). Such a balancing of complementary or opposing rationalities is compensated for, imply Feldman and March (1981), by the collection of excessive information. Such an excess of information is never really put to use, but collected in order to signal that the members of the decision-making community have done all they could be expected to do to ensure a qualified decision is made on the solid foundation of the accessible data and information. This is apparently a practice that is inherently wasteful, but it would be fair to say it is a form of waste without a purpose, a "pure squandering" of resources. Most students of decision-making would not suggest that this is the case. Instead, the very nature of decision-making is based on the notion of excessive data and information; at times, to the point where it is practically impossible for individuals to cognitively grasp all of the information needed to make adequate decisions. Rather than being an irrational reminiscence from traditional societies, an embarrassing reminder of archaic procedures, excess lies at the very heart of contemporary life. It is here, there, and everywhere – at times conspicuous and debated, but mostly infrastructural and taken-for-granted.

The issue of executive compensation

A conspicuous example of the squandering of financial resources in organizations is the compensations paid to CEOs and other executives of organizations. Rushkoff (2009) summarizes the changes during recent decades:

> [T]he average CEO's salary [corresponded to 179 times the average worker's pay in 2005, up to from a multiple of 90 in 1994. Adjusted for inflation, the average worker's pay rose by only 8 percent from 1996 to 2005; median pay for chief executives rose 150 percent. The top tenth of 1 percent of earners in America today make about four times what they did in 1980. In contrast, the median wage in America (adjusted for inflation) was lower in 2008 than it was in 1980. The number of 'severely poor Americans' – defined as a family of four earning less than $9,903 per year – grew 26 percent between 2000 and 2005. (Rushkoff, 2009: 181)

Khurana (2002), studying the recruitment process for a new "charismatic CEO" at an American investment bank, also points to the significant growth in CEO compensation:

> One study that examined CEO pay levels at publically held corporations found that CEO pay jumped 535 percent in the 1990s, dwarfing the 297 percent rise in the S&P [Standard&Poor] 500, a 116 percent rise in corporate profits, and a 32 percent increase in average worker pay (not adjusted for inflation) … If the minimum wage had risen as fast as CEO pay, it would now be $24.13 an hour instead of $5.15, which is less, in real dollars, than it was in 1970. (Khurana, 2002: 191)

According to trade union organization AFL-CIO, the CEO to "average full-time US worker" pay ratio has changed from 42: 1 in 1980 to 107: 1 in 1990, reaching an all-time-high ratio in 2000 at 525: 1 – primarily an effect of the bull market on the American stock exchange of the "new economy" financial bubble – while falling to 319: 1 in 2008, but still close to three times higher than in 1990 (Lazonick, 2010: 699). AFL-CIO's "CEO paywatch" homepage reveals that, in 2009, the total average compensation for American CEOs was USD 9.25 million, a figure calculated using public pay data from 292 companies in the Standard & Poor 500 index. This sum of USD 9.25 million includes a number of items such as salary (USD 1.04

million), bonus (USD 203,714), stock awards (USD 2.63 million), option awards (USD 2.28 million), non-equity incentive plan compensation (USD 1.79 million), pension and deferred compensation earnings (USD 1.06 million), and finally, the residual item "all other compensation" (USD 235,232) (AFL-CIO Executive paywatch). Table 1 summarizes how this economic compensation paid to CEOs relates to the average salary in US industry.

Table 1: Development of CEO pay and compensation in the US, 1980–2009, (Source, AFL-CIO Executive paywatch, URL: http://www.aflcio.org/corporatewatch/paywatch/pay/, Accessed April 7, 2011)

Year	Times average salary
1980	42
1990	107
2000	525
2006	364
2007	344
2008	319
2009	263

Lazonick and O'Sullivan (2000: 25, Figure 5) show that, between 1980 and 1995, factory wages went up 70 percent, 15 percentage units lower than the inflation of 85 percent, leading to reduced real wages for factory workers during the period. During the same period, corporate profits increased by 145 percent and CEO compensation by 499 percent. In the UK, a similar increase in CEO compensation has been reported: "At the turn of the millennium, top bosses took home forty-seven times the average worker's wage. By 2008, they were earning ninety-four times more" (Jones, 2011: 163). In addition to salary, equity-based compensation (stocks, options, and other financial instruments, commonly referred to using glaring metaphors such as "golden parachutes" and "golden handcuffs") has grown considerably during the last two decades. "The value of CEO equity portfolios has typically been more than five times higher than their annual compensation", remark Lord and Saito (2010: 53). In addition, generous bonuses have become a standard form of compensation in many industries. For instance, in the finance industry, enormous amounts of money have changed hands, even during periods of economic turmoil. Despite the economic difficulties, Wall Street bonuses amounted to USD 32 billion in January 2008, only a fraction less than a year earlier. In 2007, the finance industry in the US earned USD 53 billion in overall compensation while Goldman Sachs alone, ranked at the top of the five leading brokerages, accounted for USD 20 billion of that total, thus paying out more than USD 661,000 in wages and bonuses per employee. The CEO of Goldman Sachs alone, Lloyd Blankfein; took home USD 68 million (Sorkin, 2007: 4).

In a rational choice theory model explaining this growth in compensation, it is the shareholders, the owners of the company, who make the informed decision that increased compensation is beneficial to their economic interests, that a well-compensated CEO is capable of performing better than a less-well compensated CEO. Agency theory advocated by, for example, Jensen and Meckling (1976), conceives of the firm as a bundle of contracts (see also Berle and Means's seminal work [1991] and Alchian and Demsetz, [1972]): "It is important to recognize that mostly organizations are simply *legal fictions which serve as a nexus for a set of contracting relationships among individuals*", write Jensen and Meckling (1976: 310). The central mechanism of agency theory is that one or more persons, the so-called *principal(s)*, engage another person – the *agent* – to "[p]erform some service on their behalf which involves delegating some decision making authority to the agent" (Jensen and Meckling, 1976: 308). The principal(s) and the agents articulate and formalize their responsibilities and expectations in a contract that regulates their relationship. According to agency theory, compensation, i.e., the "managerial pay package", is "probably the most potent tool available to the stakeholders to align the interest of the managers with their own" (Lord and Saito, 2010: 40).[26] So, assuming that the principals, i.e. the shareholders, expect a linear relationship between compensation and performance, or at least a positive correlation – which would be consonant with the *homo oeconomicus* model, prescribing that, all things being equal, economic actors prefer more over less (e.g., higher interest rates on a bank account) – one can expect the increase in pay to be in parity with performance. If this is not the case, agency theory has either been based on the wrong premises or the principals have consistently overstated the causality between compensation and performance, i.e., either the theory has been falsified or the actors (principals) have been behaving irrationally (i.e., they have been behaving differently than postulated by the theory).

---

26 In an article published in the *Harvard Business Review* in 1990, Jensen and Murphy argue that American corporations offer their CEOs and executives compensation that is too low and of the wrong kind. Such a financial undervaluing of their competencies leads, argue Jensen and Murphy (1990: 138) in the characteristic rhetoric of the proponents of agency theory, to the loss of opportunities: "On average, corporate America pays its most important leaders like bureaucrats. Is it any wonder then that so many CEOs act like bureaucrats rather than the value-maximizing entrepreneurs companies need to enhance their standing in world markets?" Instead, Jensen and Murphy (1990) suggest that pay should be more closely related to "performance" because "highly talented people who would succeed in any field are likely to shun the corporate sector, where pay and performance are weakly related, in favor of organizations where pay is more strongly related to performance – and prospects of big financial rewards more favorable" (Jensen and Murphy, 1990: 149). Apparently, Jensen and Murphy (1990) struck a chord in corporate America as executive compensation has both soared and been restructured post-1990.

"The median levels of total real annual CEO compensation more than doubled from \$1.18M in 1994 to \$2.80M in 2007 (in real 1994 dollars)", report Lord and Saito (2010: 43). During the 1990s, compensation in equity increased sharply, especially during the 1999–2001 period, prior to the "New Economy" bubble; however, since 2003, "real total compensation has risen again significantly in most industries" (Lord and Saito, 2010: 43). Based on these statistics, it would be fair to say that CEOs are well compensated for their efforts. The question is, then, whether or not this growth in compensation, from 42 times the average salary to 263 times the average salary, is justified on the basis of economic reason (putting moral issues regarding the distribution of wealth in society aside), i.e., whether or not this increase in pay is based on informed decisions and rests on solid evidence?

Bebchuk and Grinstein (2005: 286) studied CEO pay during the 1993–2003 period and found, based on statistical methods and regression analyses, that "[c]ompensation levels increased far beyond what can be attributed to changes in size and performance". When pay was adjusted to changes in performance and corporation size, only 40 percent of the rise in pay could be explained: "[C]hanges in size and performance can explain only 66 per cent of the total 166 per cent increase, or about 40 per cent of the total increase, with 60 percent of the total increase remaining unexplained" (Bebchuk and Grinstein, 2005: 287). During the 1993–2003 period, CEO pay increased according to the predicted 1993 regression by 115 percent, i.e., CEOs more than doubled their compensation. Top-five executives increased their compensation by 79 percent during the same period. These findings were statistically significant at the 1 percent level, report Bebchuk and Grinstein (2005: 289). "We thus conclude that the relationship between pay and firm attributes has changed substantially during the period under consideration", write Bebchuk and Grinstein (2005: 289). Regarding equity-based compensation, when checking for changes in firm size and performance, CEOs increased their compensation by a factor of 1.347, and top-five executives by a factor of 1.468 (Bebchuk and Grinstein, 2005: 291), while there has been a decline in equity-based compensation post-2000, "[c]ash-based compensation trended upwards throughout the examined period, and its growth pace even picked up after 2000" (Bebchuk and Grinstein, 2005: 291). Finally, "although equity-based compensation has grown the most, its growth has not been accompanied by a reduction in cash compensation", note Bebchuk and Grinstein (2005: 302). Bebchuk and Grinstein (2005) thus provide robust evidence that, during the 1993–2003 period, CEOs doubled their compensation while top-five executives raised theirs by 79 percent *ceteris paribus*. Sixty percent of the total increased compensation "remains unexplained". Khurana (2002: 191) notes that this unprecedented and irrational growth – given the proposition in neoliberal economic theory that economic markets price resources correctly on the basis of their value (i.e., so-called "market efficiency") – in economic compensation paid to CEOs, uncorrelated to actual performance, is one of the most salient effects of investor capitalism: "Golden parachutes, golden handcuffs, and the whole panoply of mechanisms for lavishly rewarding CEOs without regard to performance – all of which were unheard of before the age of investor capitalism – became standard features of CEO pay packages." Khurana's (2002) explanation for this systematic overcompensation is a combination of what have been called "small-number markets" – there being, despite the substantial economic compensation involved, an endemic shortage of qualified candidates – and the board of directors' cognitive limitations in terms of ignoring or seriously underestimating the risks attached to "outside succession", i.e. the recruitment of CEOs from outside the firm. Ultimately, Khurana (2002) suggests, in investor capitalism, firms no longer acquire or maintain their own stock of capital – shareholder value ideologies prescribe that profits be given to shareholders – but increasingly rely on the financial markets (see, for instance, Ho, 2009) and, consequently, firms need a "charismatic" CEO capable of communicating with external financial analysts. The growth in CEO compensation is, consequently, a matter of deep-seated institutional changes in the capitalist economic systems. Joseph E. Stiglitz (2010b) suggests that malfunctioning corporate governance is part of the explanation:

> American corporations (and those in many other countries) are only nominally run by the shareholders. In practice, to a very large extent, they are run by and for the benefit of management. In many corporations where ownership is widely diversified among disparate shareholders, management effectively appoints most of the board, and it naturally appoints people who are likely to serve their interests most effectively. The board decides on the pay of management, and the 'company' provides good rewards for its board members. It's a cozy relationship. (Stiglitz, 2010b: 154)

While there are behavioral theories explaining why boards of directors and managements act as they do, Stiglitz (2010b) is more concerned about the shareholders not paying attention to such deficient practices, being so obviously in conflict with the predominant shareholder value ideology and mainstream economic theory: "That the executives had the incentive – and the tools – to design compensation packages that benefitted them at the expense of others seems abundantly clear. What still is a mystery is why shareholders didn't recognize this." (Stiglitz, 2010b: 154) Worse still, as Lazonick (2010: 700) claims, "U.S.-style stock-based executive compensation separates the interest of those who exercise strategic control from the rest of the corporate organization. It also subverts the process of organizational integration by sacrificing the interest of employees for the sake of booking profits. And it reduces the financial commitment to investment in innovation." Excessive (i.e., above and beyond what is justified by performance) executive compensation thus undermines – contrary to what has been asserted by, for instance, Jensen and Murphy (1990) – rather than reinforces sustainable competitive advantage. The agency theory view used to justify excessive executive compensation has, says Perrow (1986: 18), a "very conservative political bias". In addition, agency theory is based on assumptions regarding "human behavior" and organizations that are, suggests Perrow (1986: 11), "not only wrong but also dangerous".

Agency theory suggests that the principal makes informed decisions justified using data, solid evidence, and rational arguments (Jensen, 1993). In the case of CEO compensation, not even half of the increase in pay can be justified on the basis of such rational arguments, solid evidence and data. In order to save the theory, potentially with the *ad hoc* hypothesis that principals operate under the influence of bounded rationality, the question is, then, why these intentionally rational agents have failed to learn from experience over a period of 20 years (in Bebchuk and Grinstein's 2005 study)? Time and again, pay in excess of performance and other explanatory factors has been awarded to the CEO. Agency theory, embedded in the rational choice theory and the ideal-typical *homo oeconomicus* model, predicts that the contract between the principal and the agent safeguards the interests of the principal, in this case the shareholders. If this is not the case, there will be little room for any other explanation than this being a form of squandering.

Squandering and innovation: The importance of organizational slack

Slack is the term used for the surplus resources that organizations acculumate, i.e. that which is in excess of the absolute minimum of resources required in order to maintain organizational activities and functions (March and Simon, 1958; Cyert and March, 1963; Näslund, 1964). Nohria and Gulati's (1996) much-cited paper offers a comprehensive definition:

> We define slack as the pool of resources in an organization that is in excess of the minimum necessary to produce a given level of organizational output. Slack resources include excess inputs such as redundant employees, unused capacity, and unnecessary capital expenditures. They also include unexploited opportunities to increase outputs, such as increases in the margins and revenues that might be derived from customers and innovations that might push a firm closer to the technology frontier. (Nohria and Gulati, 1996: 1246)

Slack has commonly been defined as some kind of "safety deposit", some additional extra resources ready to be put to use whenever needed. However, over time, and in the face of the increased emphasis on new corporate governance routines and institutional logics, slack has become, say Floyd and Woolridge (1994: 54), a "dirty word". At the same time, they continue, "flexibility, experimentation, and learning" do require resources (Lawson, 2001). A few studies have been concerned with the relationship between innovation and slack. Nohria and Gulati's (1996) study proposed an inverted U-shaped curve where innovativeness increased with slack up to a certain point where too much slack then reduced the attention to innovativeness. Geiger and Cashen (2002) further developed the research methodology by using multivariate methods and showed, just like Nohria and Gulati (1996), an inverted U-shaped relationship between innovation and slack; too little slack reduces the proclivity to take risks, thus hampering innovativeness, while too much slack undermines the discipline of the co-workers. In a more recent study, Love and Nohria (2005) make a distinction between *available slack*, e.g. retained earnings, and *absorbed slack*, which is "not easily redeployed". Excess personnel is one example of absorbed slack. As Love and Nohria (2005: 1090) remark, agency theorists tend to identify slack with "inefficiency" – the proverbial excess "fat" of the corporation that can be spared – and that "slack is accumulated and misused because of principal-agent problems", especially when managerial tasks are complicated to monitor fully. When it is difficult to control output, agency theorists postulate that managers accumulate "excess resources". However, despite such persistent critique, Love and Nohria (2005: 1089) suggest that "[s]lack itself has received little attention as an important contingency, notwithstanding frequent references to 'fat' in the rhetoric around downsizing". In their empirical study of the 100 largest American companies between 1977 and 1993, Love and Nohria (2005) again found support for the inverted U-shape curve between slack and performance, but they also found that firms that are downsizing – cutting down on slack – should take a broad and proactive approach to achieving improved performance: "Overall, downsizing in our sample of large industrial firms had no main effect on subsequent firm performance … High absorbed slack firms that downsized proactively using a broad scope approach were most likely to achieve performance improvements" (Love and Nohria, 2005: 1103). They continue: "Our results support the intuitive proposition that downsizing is more likely to lead to improved performance when firms are 'fat' or have higher degrees of absorbed slack" (Love and Nohria, 2005: 1103). At the same time, Love and Nohria (2005) warn against general and sweeping criticism regarding fat in organizations since this kind of criticism fails to take into account contingencies and local conditions. Contrary to agency theorists' claims, Love and Nohria (2005) reject the idea that slack is unproductively used resources; instead, in fact, slack leads to increased performance and innovation up to a certain point, as prescribed by the inverted U-shape model. Such findings are also supported by the study of Greenley and Oktemgil (1998), who emphasize, in their study of 134 British companies in 21 industrial sectors, that high-performing companies accumulated a higher degree of slack than did low-performing ones. Mellanhi and Wilkinson (2010) use a more sophisticated, longitudinal method to examine how downsizing affects the innovative capacities of the firm over time. They found that downsizing has "a large impact on innovation" but that the magnitude of the impact "declines over time" as the firm adjusts to a new equilibrium (Mellanhi and Wilkinson, 2010: 488). In providing this evidence, the inverted U-shape relationship was somewhat modified as a low degree of downsizing did not affect the innovation rate very much:

> [L]ow downsizing – 5–8 percent – had a very marginal positive effect on innovation. However, large downsizing had a significant negative impact on innovation output. Therefore the relationship between level of downsizing and innovation two years post-downsizing is more of an inverted '$\sqrt{}$' than a 'U'-shaped relationship. (Mellanhi and Wilkinson, 2010: 497)

Mellanhi and Wilkinson (2010) suggest that "excessive downsizing" has a significant effect on innovation, but that the impact is temporary. Interviews held at companies undergoing restructuring reveal why innovation is negatively affected. First, the allocation of resources to innovation activities decreases, thus starving innovation projects. Second, human resource slack becomes depleted, leading to a lower level of risk-taking and degree of experimentation. Third, ongoing innovation processes are disturbed as new concerns and issues need to be dealt with on a short-term basis. Fourth, and finally, "employment relations deteriorate, creating an environment that is not conducive to innovation" (Mellanhi and Wilkinson, 2010: 501).

As major downsizing activities negatively affect innovation, Love and Nohria (2005) as well as Dougherty and Bowman (1995) provide a number of recommendations to managers regarding how to maintain innovative capacities during and after downsizing. Besides preventing a loss of risk-taking that is too significant, Dougherty and Bowman (1995), who also find a negative relationship between downsizing and innovation, emphasize the need to maintain the human resource systems and networks that are the life-blood of any innovative firm. If the informal networks and communities within a firm are eliminated, it may take years to re-construct such human resources. At the same time, as Bowen (2002) emphasizes in his study of the relationship between the greening of industry initiatives and slack, one must not assume that the presence of slack *per se* will inevitably lead to any innovations, especially any radical innovations such as more sustainable products and production systems:

> Organizational slack seems an important factor in corporate greening, but it is unlikely that forms with relatively higher levels of slack will spontaneously become more environmentally responsive … Slack may facilitate strategic environmental behaviours but will not necessarily initiate them. (Bowen, 2002: 315)

Slack is thus a resource that needs to be managed properly in order to be translated into durable effects such as innovations.

Research on the relationship between slack and innovation has been complemented by the study of Herold, Jayaraman, and Narayanaswamy (2006) which operationalizes slack as "liquidity", i.e. the ratio between "current assets, minus inventories, divided by the current liabilities" (Herold, Jayaraman, and Narayanaswamy, 2006: 381). Herold, Jayaraman, and Narayanaswamy (2006) measure innovative capacities as the degree of patenting, the amount of patent applications filed. Studying 261 companies over a nine-year period (1990–1999), Herold, Jayaraman, and Narayanaswamy (2006: 384) found, largely consonant with the study of Nohria and Gulati (1996), that "too little slack may inhibit exploration programs that lead to innovation, while too much slack may, in fact, result in reduced benefits beyond a certain point". Slack is thus beneficial to innovation within certain limits and under determinate conditions. Taking a different approach to studying slack and innovation, Richtnér and Åhlström (2010) argue that studies of "financial slack" may be illuminating, but they do not "[n]ot address the nature of the relationship between organization slack and innovation at the level of the new product development project" (Richtnér and Åhlström, 2010: 376), i.e., the sites where innovations are produced within organizations. Richtnér and Åhlström (2006: 429) consequently define slack, on the level of project development projects, as "the possibility to deviate both from agreed upon deadlines and from original project specifications" (Richtnér and Åhlström, 2006: 429), that is, slack is defined in temporal rather than financial terms. Based on their empirical studies of "high velocity industries", Richtnér and Åhlström (2006: 436) found that the absence of slack in terms of project deliverables "[s]eems to reduce the possibilities for a focus on tacit knowledge, which hampers the ability to create knowledge and ultimately innovation". More specifically, Richtnér and Åhlström (2006: 436) observed that projects with less slack tend to focus more on explicit rather than tacit knowledge and to rely on formal procedures and deliverables such as stage-gate models and deliverables when managing project work. In Richtnér and Åhlström's (2006) view, the focus on explicit knowledge may affect the innovative capacities of the firm since findings and insights not anticipated by formal project management models may be excluded or overlooked. Richtnér and Åhlström (2010: 400–401) are critical of the concept of slack as a unidimensional and general concept, identifying four categories of slack in product innovation projects: (1) *project deliverables slack*, the ability to depart from deadlines and/or product functionality to explore new possibilities; (2) *human resource slack*, the "level of collective competence of the project members"; (3) *consumer interaction slack*, "the interaction between a project and its customers"; (4) *top management control slack*, "the control that top management exerts over a project". Richtnér and Åhlström (2010) suggest that, depending on what slack is being manipulated, different outcomes can be expected and thus one managerial implication is that managers need to be aware of how slack influences knowledge creation as well as sharing and innovation within the company.

Regardless of Richtnér and Åhlström's (2010) classification of more detailed slack categories, slack remains a somewhat complicated term in the management literature; formally speaking, it is "unproductive capital" while apparently still playing a role in creating innovative thinking. In general, slack is not defined in the same way in which Bataille speaks of squandering, the ritual *destruction* of resources, but it remains the resources that are *excessive*, that are never put to productive use in the conventional sense of the term. In a regime of corporate governance preoccupied with the circulation of invested capital, slack is, consequently, a form of what Plato in *Phaedrus* speaks of as the *pharmakon*, the medicine/poison/magical potion that is both good and harmful at the same time (see Derrida, 1981); on the one hand, it is something that needs to be eliminated in order to safeguard the effective use of resources, while on the other, it is something that has the potential to create new innovations leading to new income and the long-term survival of the firm. Slack haunts managers as it is complicated to fully monitor and assess in terms of its function. In the same way fat accumulated over time in the human body (see Näslund [1964] for this biological metaphor) may be regarded as energy stored for later use (in, for example, the case of famine, evolutionarily speaking a not too distant experience for humans, either in time or in space), these excessive resources are not part of the operating capital and need to be eliminated or put to use where there is more pay-back. Still, studies of slack as a precursor to innovation, or as an element of the innovative capacity, demonstrate that there is, within certain limitations, a positive relationship between, on the one hand, innovative capacity and, on the other hand, slack. The saying "necessity is the mother of invention" is thus misleading since necessity implies a sense of scarcity and short-term concern; instead, invention is produced on the basis of excess resources, when time and energy can be released for productive use. Only after accumulating adequate resources, a reservoir of time and energy for creative thinking, will social communities be able to develop new ideas and new thinking (Leroi-Gourhan, 1993). This applies to both mankind in general and organizations – *nihil ex nihilo fit*, out of nothing, nothing comes. Slack must thus be understood, not so much as irrationally reminiscent of poor management and sloppy householding (as postulated by, for instance, Jensen, 1993) but as the basis for innovative thinking and innovation work, the minimal excessive resources needed to produce new ideas and novelty. Given the recent decades' emphasis on shareholder value creation, increased pay-out ratios, and the increased compensation paid to CEOs and top management, in short the financialization of corporate governance (e.g., Bebchuk and Grinstein, 2005; Krippner, 2005; Lazonick and O'Sullivan, 2000), the lack of innovativeness may potentially be explained on the left-hand side of the documented inverted U-shaped curve, i.e., the draining of productive resources from firms and industrial corporations has gradually undermined the capacity to innovate.[27] Seen in such a perspective, slack is not an affront to economic reason but rather an investment

---

27 Froud *et al.* (2000) argue that the emphasis on shareholder value does not, unlike claims made by its proponents stressing the immediate financial rewards received by entrepreneurs, promote innovations in industry. Using the case of the Japanese automotive industry, a source of much acclaim in the 1980s when, for instance, the large-scale international automotive industry research project leading to the best-seller *The Machine That Changed The World* (Womack, J., Jones, D.T., and Roos, 1990), the book that coined the phrase *lean production* (today more widely referred to as *lean*) portrayed the Japanese manufacturing industry as more or less an industrial and managerial miracle. Despite implementing a number of new practices and routines undoubtedly leading to, for example, more effective logistics (so-called *just-in-time logistics*), Japanese industry was heavily favored by a variety of macro-economic factors and structural advantages, argue Froud *et al.* (2000: 106), which included "high utilization, long hours, favourable exchange rates and the resulting low dollar wages" that for the most part were overlooked or underestimated by the analysts of Japanese manufacturing industry. "The Japanese could never build a car in half of the hours of a US assembler, as Womack *et al.* (1990) contended", claim Froud *et al.* (2000: 106).

Similarly, the more recent praise of shareholder-value-based corporate governance rests on a similar inability to examine the whole macroeconomic system and the structure of the economy: "The homiletic examples in the consultancy literature obscure the central point that management actions and corporate positions that work to deliver shareholder value for some firms will not work for all. Corporate successes within one sector are in some cases achieved at the expense of failures elsewhere; and in more cases, the management actions or performance of successful firms rest on special case advantages which cannot be copied or will not produce equally large benefits for all who imitate" (Froud, Haslam, Johal, and Williams, 2000: 108). Froud *et al.* (2000) thus argue that shareholder-value-based corporate governance is based on a fallacy; its proponents point to the release of capital that can be productively invested in new ventures (see, for instance, Jensen [1993] for a firm defense of such a view). However, for Froud *et al.* (2000), such capital is only concentrated to certain groups (a claim that, for example, Duménil and Lévy [2004] and many other commentators have proven to be correct), with little incentive to take further risks as they already control substantial amounts of capital. In other words, while proponents of shareholder-value-based corporate governance assume that capital is piped into enterprising and entrepreneurial communities (as is actually the case in certain clusters and regions, including Silicon Valley. See, for instance, Ferray and Granovetter, 2009; Saxenian, 1994), skeptics like Froud *et al.* (2000) say that the capital previously used to promote innovations in-house is more likely to end up in pension funds and in the hands of the one percent of the population who own a vast proportion of the accumulated economic resources.

in the resources from which novelty can be produced. Potentially, new and more innovation-friendly corporate governance policies may gain a foothold in industry and in the management writing and economic theory of the future.

Summary and conclusions

Of the three principles of situational rationality and economic agency discussed in this book, the principle of squandering is, perhaps, the most counter-intuitive, underlining the fact that there is a continuous and ongoing expenditure of resources which, in many ways, is at odds with the master narrative of capitalism, i.e. that there is a scarcity of resources and that all assets need to be used wisely. Underlying the economy of scarcity, the restricted economy, is a general economy of abundance. When it comes to bonuses in the finance industry, and the salaries and benefits of CEOs and board members, different principles apply than in cases where, for example, the cost of production is monitored and evaluated. This does not suggest that CEOs and other decision-makers in organizations are cynically seeking to maximize their economic compensation at the expense of other social interests; instead, it demonstrates that agents in control of the economic system of competitive capitalism have developed a firm belief in providing conspicuous economic compensation for *some* competencies and skills while grossly underrating others. *Belief* rather than *greed* is the motor of the uneven distribution of resources in contemporary society; while greed is short-sighted and clumsy, belief is noble and self-righteous. For instance, the financial performance of an enterprise is dependent on a long series of contingencies and macroeconomic factors, but still the CEO gets the credit for the financial performance reported. Drawing on the works of Georges Bataille, the substantial amounts spent on CEO salaries and benefits are indicative of how the surplus must be squandered rather than reinvested in the firm or transferred, as higher salaries, to the employees. A calculative mindset, seeking to maximize the very use of economic resources within a firm or industry is helpful when it comes to leveraging the return on investment, but fails to identify or create new domains of investment. Therefore, the restricted economy of calculation and accounting and the general economy of expenditure operate in parallel in many organizations and economic fields. While costs have to be cut in, for example, public sector healthcare, there may also be room for more venturesome investment in uncertain domains. Again, similar to the case of reciprocity, competitive capitalism is paradoxical inasmuch as some fields are characterized by scarcity and frugality while others are endowed with substantial resources and conspicuous consumption. Exactly which domains end up in both positions is a matter of politics and belief, rather than strictly calculative evaluations. The concept of the general economy proposed by Bataille is helpful in theorizing what seems to be one of the great enigmas of contemporary capitalism, i.e. how certain competencies may be worth so much more than other skills and know-how. The answer is that, rather than being based on formal calculation and empirical data, the economic compensation paid to certain experts in the management of capital is based on the principle of expenditure, the waste of economic resources, since a reinvestment of the surplus would lead to other undesirable effects such as inflation or diminishing returns on investment due to higher labor costs. Like no other social theorist, Bataille emphasizes how non-utilitarian rationalities, in many cases ignored, overlooked, or explicitly refuted, serve both as the underside of the capitalist economic systems and in pre-modern economies. Economic agency is not only a matter of maximizing output; under determinate conditions, the waste and squandering of economic resources serves to release surplus economic resources. While the contemporary master narrative of competitive capitalism rejects such non-utilitarian rationalities as mere nonsense, there are still only a few credible explanations as to why the massive economic surpluses generated consistently fail to be put to use where they are desperately needed; instead, issues regarding, for example, conspicuous economic compensations are justified on the basis of overtly rationalist arguments that speak of "competence" and "experience" being highly valued. However, such arguments fail to explain, among other things, why bonuses are paid even during periods of loss, or when stocks are plummeting; thus, Bataille's argument regarding the squandering of the accursed share offers an alternative story to be told.

Chapter 5.

A social theory of innovation

Introduction

"Criticality is … not a position per se, not a site or a place
that might be located within an already delimitable field,
although one must, in an obligatory cathesis, speak of sites,
of fields, of domains. One critical function is to scrutinize
the action of delimitation itself."

Judith Butler (2004: 107)

"All thinking in categories, all nascent thinking in schemata, that is, in accordance with rules, is *perspectival*, conditioned by the essence of life", suggests Heidegger (1987: 102) in his massive three-volume work on Nietzsche, underlining the notion that all human knowledge is situated, contingent, contextual, in short, using Nietzsche's term, *perspectival*. Knowledge does not fall from the sky but is produced within certain communities and under specific social conditions and, consequently, the providers of such knowledge are not in the position to speak on behalf of their knowledge "from above" or from some external position outside of society. Knowledge production is, for good and bad, a part of this world, "co-produced" as Jasanoff (2005) puts it, with society at large. As a consequence, knowledge, forms of knowing, and modes of thinking are perspectival, embedded in the communities of expertise and shared professional interests (Fleck, 1979). As Butler (2004) emphasizes, critique is something that occurs *in medias res*, in the middle of things, here and now, where we stand, as we speak. Lemke (2011: 29) suggests that Michel Foucault distinguished between a "juridico-discursive" form of critique, a style of thought "focusing on judging and condemning, negating and rejecting," and thus implicitly demanding the "determination of rational standards of evaluation and the application of those standards to social reality". In Lemke's (2011: 29) account, "this form of critique appears to be a predominantly negative practice characterized by deficit, dependency, and distance". In contrast to this juridico-discursive critique, Foucault positions a form of critique not based on "a solitary attitude" but on that which is "more local and 'experimental'" and more closely connected to existing forms of government. This critique does not primarily judge, but examines and discusses. It participates rather than distances. Such a form of critique is important for the social sciences and for the study of organizing and managerial practice since it enables a critical view without becoming distanced from the object of study.

The "thinking in categories" that are trained and disciplined in professional communities are conventions which, in various ways, embody particular professional ideologies and aspirations, but also wider social interests and concerns (see, for example, Latour, 1988); no forms of knowing the world are able to subsist without the legitimacy of the larger social community in which they are advanced (Shostak and Conrad, 2008). In addition, since all societies produce a repertoire of legitimate, semi-legitimate, and illegitimate modes of knowing and thinking, there is a continuous struggle for legitimacy between various worldviews and perspectives. In the US, the Christian right is struggling to disqualify Darwinian evolutionism because such doctrines violate their fundamentalist reading of the Bible, and responds by proposing alternative analytical frameworks such as creationism and intelligent design. In the field of medicine, various forms of so-called "alternative medicines" are facing harsh criticism from the community of mainstream physicians and physiologists because these therapies rarely rest on the adequate clinical evidence demanded during the contemporary period of scientific medicine (Collins and Pinch, 2005; Bynum, 1994). Various perspectival regimes of knowledge thus complement one another and struggle to influence the general mindset. When it comes to economic agency, the actions within the specific domain of industry and enterprising activities, there is a strong emphasis on instrumental rationalities, modes of thinking that are relatively recent articulations of a particular form of instrumental reason privileging certain forms of thinking. In an economic regime characterized by scarcity, such instrumental reasoning may serve the function of directing attention toward issues needing to be resolved; however, in the late modern regime of competitive capitalism, it may be the case that alternative rationalities could play an increasingly important role. Late modern society and its economy are characterized by an overflow of information and know-how and, in order to navigate in this domain and make the best use of various resources and opportunities, alternative perspectives may need to be taken. This book has served to advance one or two such complementary perspectives which, in various ways, could inform the study of organizations and managerial practice.

In this final chapter, some of the loose threads and unfinished discussions are rearticulated in order to point to the value of broadening the scope and perspective regarding how organization could be both enacted and understood in the new economic regime, characterized by an abundance of information and know-how and desperately calling for creativity, entrepreneurial skills, and innovative thinking. Today, only entrepreneurs can save us, claim politicians and policy-makers. During the contemporary period, economic value is not merely generated from the engineering of nature into commodities, but also on the basis of aesthetic, stylistic, and creative competencies (Townley, Beech, and McKinlay, 2009; Koivunen and Rehn, 2009; Caves, 2000). This shift in focus from material transformation to aesthetic creation may preoccupy business school researchers during the forthcoming period. In order to fully recognize and appreciate this shift in perspective, there is a need to reevaluate the very concept of rationality at its root.

This chapter is structured accordingly: First, the concept of mechanical clock-time will be discussed as the *primus motor* of competitive capitalism during the modern period, instituting the seriality of equidistantial time serving as the foundation for the *homo oeconomicus* model. Thereafter, a few studies of how temporality is enacted within organizations will be reported on. Third, the three principles will be revisited, examined one more time as part of the situational rationality and economic agency that have been relatively marginalized in organization theory and management studies. Finally, some implications for organization studies will be addressed. The chapter finishes with a summary of the key arguments of this volume.

Tempus vitam regit: Instrumental rationality and the mechanical clock-time of modernity

The British anthropologist Edward Evan Evans-Pritchard, part of the first generation of anthropologists in the service of the British Empire, is best known for his study of the Nuer, a people organized as a confederation of tribes living in Southern Sudan and Western Ethiopia. Among the many interesting things that Evans-Pritchard observed in the Nuer society, the absence of the time *qua* analytical category constitutes a striking difference in relation to European societies:

> The Nuer have no expression equivalent of time in our language, and they cannot, therefore, as we can, speak of time as though it were something actual, which passes, can be wasted, can be saved, and so forth … Events follow in a logical order, but they are not controlled by

an abstract system, there being no autonomous points of reference to which activities have to conform with precision. Nuer are fortunate. (Evans-Pritchard, cited in Thompson, 1967: 96)

For the Nuer, time is what Bergson calls *durée*, duration, lived time, time that is embodied, part of everyday existence, rather than being the mechanical clock-time of equidistantial time units ceaselessly adding one time-unit to the other in an eternal sequence leading from the past into the future through the transient passage point of the present. The Nuer do not believe one can "control time"; time simply *is*, outside of any "autonomous point of reference". The Nuer are indeed fortunate.

Lewis Mumford (1934) remarked that the most important capitalist technology is the clock, enabling the coordination of tempospatially diverse activities. "Standard time is … among the most essential coordinates of intersubjective realitiy, one of the major parameters of the social world. Indeed, social life would probably not have been possible at all were it not for our ability to relate to time in a standard fashion", Zerubavel (1982: 2) assures us. Clock-time and mechanical time-keeping devices were invented in the monasteries of the fourteenth century to ensure prayers and ceremonies were timely (Adam, 2006: 125, see also Gurevich, 1964; Landes, 1983); however, outside the confined sphere of the monasteries, it took centuries before time-keeping became widespread. Agrarian production is relatively uncoordinated and thus in relatively little need of synchronicity. It was not until manufacturing became concentrated to factories and workshops at the end of the eighteenth century that a more substantial need arose for coordinated time-keeping. Remaining in this setting; the British historian E.P. Thompson (1967) suggests that there was relatively limited demand for controlling workers using temporal schedules:

> Attention to time in labour depends in large degree upon the need for the synchronization of labour. But in so far as manufacturing industry remained conducted upon a domestic or small workshop scale, without intricate subdivision of processes, the degree of synchronization demanded was slight, and task orientation was still prevalent. (Thompson, 1967: 70–71)

In line with Max Weber's (1992) well-known thesis that protestant religious beliefs advanced capitalism, Thompson (1967) argues that Puritanism was concerned with the use of the time endowed by God and called for the careful use of time: "Puritanism, in its marriage of convenience with industrial capitalism, was the agent which converted men to new valuations of time; which taught children, even in their infancy, to improve each shining hour; and which saturated men's mind with the equation, time is money" (Thompson, 1967: 95). Others have emphasized more practical demands and needs emerging during the nineteenth century as the principal driver of the standardization of time. The railways, both in Europe and North-America, growing in Europe from 1.7 thousand miles in 1850 to 101.7 thousand miles in 1880 and in North America from 2.8 thousand miles in 1850 to 100.6 thousand miles in 1880 (Hobsbawm, 1975: 54), are commonly advanced as the single most important explanation of the enactment of a standardized time framework. "[W]ith the development of a fast railway network throughout Britain, communities which had previously led a rather autonomous existence gradually became interrelated parts of a single systematic whole", writes Zerubavel (1982: 7). With the shift from "natural to relational time", a variety of locally-enacted temporal regimes were brought into harmony. In the US Midwest, for instance, during the mid-nineteenth century, there were no less than "38 standards of time in Wisconsin, 27 in Michigan, 27 in Illinois, and 23 in Indiana" (Zarubavel, 1982: 8).

This familiar story, "the railroads brought us the new concept of time", has been disputed. Bartky (1989: 28) suggests that this "railroad time" was more like a group of "unrelated times" simply using the time in the major cities from which the railroads departed or arrived; thus, argues Bartky, "there was no 'system'". Instead, says Bartky (1989), it was scientific communities (e.g., meteorologists), in great need of coordinating their observations, who called for a state controlled standardized time system:

> Despite some 'originators' claims, standard time was not adopted primarily to bring order to the chaos of the railroad timetables. Indeed, the railroads did not need standard time for their operations. Rather, in the 1870s scientific pursuits requiring simultaneous observations from scattered points became important, and those needs led to proposals for federal actions in the early 1880s. In response to the pressures from scientists, railroad superintendents and managers implemented a standard time system on November 18, 1883, a system tailored to their companies train schedules. (Bartky, 1989: 25)

Regardless of which forces and interests helped to institute a standardized, regional, national, and global timeframe, a temporality based on "the use of clock time, the Gregorian calendar, and the Christian Era" (Zerubavel, 1982: 3), the very idea of conceiving of time as mechanical clock-time has, perhaps, been the single most important cultural trait of the late modern period. As Kula (1986: 18) remarks, "the right to determine measures in an attribute of authority in all advanced societies", and consequently time-keeping, has been part of the organization of society and the responsibility and prerogative of the various authorities.

As Galison (1994: 262) shows, in the nineteenth century, the era of swift capitalist global expansion (Hobsbawm, 1975), "time-keeping machines served as a cultural symbol of authority: these were the mechanisms that appeared everywhere, celebrated from literature and poetry to philosophy and political theory. According to Mayr, 'the authoritarian conception of order was directly and patently shaped by society's experience with the mechanical clock'." Landes (1983: 287) points out that the watch became a new necessity during the course of the nineteenth century: "Where world watch output at the end of the eighteenth century was around 350,000–400,000 pieces a year, it was up almost tenfold, to 2.5 million pieces, three quarter of a century later." Like no previous century, the nineteenth century made it clear that *tempus vitam regit* – "time rules life" (as the slogan of the National Association of Watch and Clock Collectors in the USA proclaims, cited in Landes, 1983: 360).

Time and authority were closely associated. Even today, the professions are commonly defined vis-à-vis other, less favored occupational groups as the category able to maintain control over their work time (Nordenflycht, 2010: 163). Consequently, it is little wonder that the professions, defined in terms of temporal autonomy, have been put forward as being in conflict with managerial procedures, preoccupied with the control of time (Raelin, 1995; Ackroyd, 1996; McGivern and Ferlie, 2007; Ackroyd and Muzio, 2007). In what Shenhav (1999) calls the managerial revolution, propelled by the new social class of engineers, the controlling of time was of key importance. Much has been written and said about Taylorism, and Merkle (1980: 292) argues that scientific management may be interpreted as a form of "secular protestant ethic, substituting output, or efficiency, for the amassing of money as the visible sign of grace". "At the end of the Progressive period [1880–1930], business philosophy was crystallized around secular engineering ideals rather than around religious, philanthropic, paternalistic, or Social Darwinist ones", adds Shenhav (1999: 36). For men like Frederick W. Taylor, waste is undoubtedly a sin in terms of being the inefficient use of resources. Merkle (1980) emphasizes here the utopian element in the ideology of scientific management; the belief in the possibility of building a society devoid of class-interests on the basis of technological rationalities:

> In an industrial society whose members were increasingly subscribing to theories of inevitable class conflict, Taylorism represented a technically based ideology. It painted a picture of a conflict-free, high-consumption utopia based on mass production; it presented techniques for the suppression of class conflict and advocated a new unity of social interest: it provided an avenue for middle-class mobility and the growth of a new professionalism … The proposals of the scientific managers became basic tenets of modern American organizational life: the de-

struction of ideologies of class warfare by the establishment of control according to neutral expertise in the hands of a nonowning, nonlaboring, professional, middle class. This was to become the American 'One Best Way'. (Merkle, 1980: 15–16)

Layton (1986: 140), too, addresses scientific management as an "extension and codification of the engineers' ideology":

> Central to this formulation was the idea of a universe composed of overarching laws of nature. The engineer saw himself as the discoverer and interpreter of these laws. They provided the basis in esoteric knowledge for a defense of autonomy and other professional values … Taylorism carried the professional aspirations of engineers from dream to concrete reality, at least in the eyes of its founders. (Layton, 1986: 140)

This move from "dream to reality" essentially relied on the control of time. Scientific management and Taylorism featured many ideas, including the notorious claim that "brain work" should be eliminated from the shop floor, but temporality still remains at the very heart of the recipe prescribed by Taylor and others. Georges Canguilhem (2008: 96) says that one of the key contributions made by Taylor, regardless of its implications and reception, was to rationalize the workers' movements so that the human organism is "aligned with the functioning of the machine". The rationalization of scientific management is, then, a form of "mechanization of the organism, inasmuch as it aims to eliminate movements that appear useless because they are seen solely from the viewpoint of output, considered as a mathematical function of certain factors" (Canguilhem, 2008: 96). Movements should create valuable output or else they are, by definition, waste (see e.g., Gilbreth, 1911). While the professional worker is a fully integrated and coherent cognitive unity, operating within a temporal horizon at least partially controlled by the individual, the worker is reduced to a number of movements strictly calculated on the basis of utilitarian principles. "Taylorism, one of the first 'scientific' approaches to productivity", writes Lefebvre (1991: 204), "reduced the body as a whole to a small number of motions subjected to strictly controlled linear determinations. A division of labour so extreme, whereby specialization extends to individual gestures, has undoubtedly had as much influence as linguistic discourse on the breaking-down of the body into a mere collection of unconnected parts." The worker is not a subject in the proper sense of the term, but is advanced, rather, as an assemblage or patchwork of practices and operations. This far-fetched reductionism has been grist for the mills of industry sociologists and management scholars for decades; however, even in the 1930s, many of the Taylorist procedures and routines appeared absurd and "the general feeling was that mechanization and systematisation had gone too far" (Shenhav, 1999: 46). Still, after modifications of the scientific management principles under the influence of the human relations movement during the interwar period, and industry sociology studies during the 1950s and 1960s, as well as other concerns regarding, for example, "organizational learning", what remains, *inter alia*, is the concern that time and temporality are of key importance to any managerial control and the monitoring of organizations.

Recent studies of perceptions of time and temporality in organizations suggest that there are more concerns regarding how to make use of the increasingly scarce resource of time, in both the work setting and the domestic sphere (Perlow, 1998; Dreher, 2003; Lingard and Francis, 2004; Watts, 2009). An alarmist subgenre addressing "overwork" has been around for more than two decades (see, for instance, Schor, 1993) and what Leslie Perlow (1999: 57) speaks of as "time famine", "[the] feeling of having too much to do and not enough time to do it", is showing no evidence of being in decline. On the contrary, many people, especially in the middle-class, are commonly expressing their concerns regarding how to balance work, family life, social obligations, and some personal interests in a way that does not lead to pathological consequences such as burnout (Löfgren, 2007). In the ideology of excessive work, notes Hochschild (1997: 113), the notion of being a "family man", traditionally seen as a mature and responsible position for serving society as a provider to one's children and a member of a civil society existing outside of the domain of work, has today "taken on negative overtones, designating a worker who isn't a serious player". Hochschild (1997: 219) speaks of "Taylorized children" as one of the consequences of such cultural shifts, meaning that family life is also becoming "ingrained with Taylorized, time pressured work practices". The very notion of spending "quality time" with one's family, conceiving of certain joint and supposedly pleasurable brief periods with one's family as a qualitative substitution for longer, but less blissful, periods of work, is a deeply ideological enactment of both temporality and family life. Contrary to this view, others argue that children want to have spatial and temporal access to their parents and being absent for most of the week is rarely compensated for by short, but more intense, sessions with them, regardless of their perceived value.

Studies of temporality in organizations

In the workplace itself, too, there is, in some industries and fields, an endemic sense of a time famine. For instance, in the healthcare sector, physicians tend to deplore their lack of time with patients and their inability to keep up with scientific advancements in their field of research, with complaints regarding increased amounts of administrative work being commonplace. In some professions, the entire industry is structured on the basis of the number of hours clients are "billed", e.g., advertising agencies and firms of lawyers. Elaine Yakura (2002a) studies what she calls *billables*, the time being inscribed into the clients' accounts, and suggests that the billing system used in, for example, consulting services institutes a uniform system for controlling performance, output, and effectiveness:

> Billing systems help to legitimate the value of consulting services by assigning uniform dollar rates to the hours billed. Billing practices ensure that the consulting services have uniform value, even where the realities belie that uniformity. They rely on the assumption that time is money. (Yakura, 2002a: 1077)

Since the consultancy industry uses bonuses to reward performance, all co-workers keep a close watch on their billables, with "all of the consultants scrutiniz[ing] their monthly reports carefully for signs of changes in the firm's performance" (Yakura, 2002a: 1090). That is, much of the internal coordination and relationships with clients outside of the firm were, in many ways, structured on the basis of the management control system based on a particular form of temporality, i.e. that of the monetized time of the billable. Yakura (2002a: 1090) says:

> Billables … allowed the consulting forms to control the workload. The billing system generated a variety of monthly reports used for different purposes. Senior management received reports that summarized individual and project activities for the entire office.

In another study reported by Yakura (2002b), the use of media to visualize time, what Yakura calls "temporal boundary objects", was studied. In project work, an organization form embedded in temporality, with a beginning and a stipulated end, such media are important in portraying time as something that is both tangible and manageable, and under the control of the project leader and other project members. In project management settings, *Gantt-charts* (a method developed by F.W. Taylor's disciple Henry L. Gantt, see, for instance, Gantt [1913/1919]), a specific form of timeline has been widely used. Such timelines, suggests Yakura (2002b: 958–959), imply a certain "narrative logic" whereby "once you start, you will finish; tasks and milestones are portrayed as flowing linearly and progressively". The timeline is helpful in graphically representing time, not in terms of clocks being circular or calendars being distributed in various ways, but in terms of being linear, normally moving from left to right in accordance with the Western tradition of reading. As a consequence, "timelines give participants an illusion of 'management' or 'control' of a project" (Yakura, 2002b: 968–969). Yakura's (2002b) study demonstrates that the image of time shared by, for example, the project teams of organizations needs to be jointly enacted and coordinated through the use of media and models, e.g. the timelines used in Gantt-charts to create a sense of being in control of time. To some extent, time *becomes* the timeline, an abstract and in many ways evocative, or even mysterious, concept that is embodied in the linear depiction of the timeline. The timeline is, argues Yakura (2002b), capable of taming the "pluritemporalism" found in organizations and project teams by imposing a shared image of temporality.

Jemielniak's (2012, 2009) study of American and Polish software engineers and computer programmers emphasizes an obsession with the time worked on the part of the managers, in the absence of other means of controlling the work done:

> To the bystander, an effective, successful programmer (or for that matter, a lawyer) looks exactly like the one who is a failure and sluggard. Only the final result of work can to some extent be measured, while the very process of labor is a black box for anyone not directly involved. (Jemielniak, 2009: 287)

This opacity of competence, the difficulties involved in visually inspecting the work conducted at software firms leads to a preference for management based on time; the "common value denominator", says Jemielniak (2009: 287), is the "recognition of time" rather than the work results: "As a consequence of placing much emphasis on time and its importance, higher than on any other inputs, its ceremonial role becomes pivotal." In this re-conceptualization of the old adage "time equals money", working overtime is seen as a contribution to the community and making a sacrifice for the company; however, perhaps surprisingly, meeting deadlines is not that important. Time spent at work and the additional time invested in the workplace are evaluated and overtime is praised. Jemielniak (2009) suggests that the community of computer programmers and software engineers, in many ways associated with the Californian-style counter-culture (Thomas, 2006; Wark, 2004), enjoys significant degrees of freedom regarding how they dress and behave at work, but they are expected to be at their desks working long hours. The community of savvy computer programmers is tamed by the clock. This emphasis on overtime is, in many cases, expected of professional elites and treated as a sign of commitment and loyalty. Ho (2009) reports in her ethnography of Wall Street investment bankers that, during his/her initial years on Wall Street, the neophyte is expected to work excessive overtime, around 110 hours of work per week, including, as with one of Ho's informants, 90 percent of weekends. Not only does Wall Street expect the best minds of each generation to work for the investment banks, it also expects them to do a significant amount of so-called "grunt work", tedious and demanding analytical work for long hours. A similar work ethos has been observed among resident doctors (Kellogg, 2011).

Non-linear temporalities

The three principles of economic agency, playfulness, reciprocity, and squandering, are all, in their idiosyncratic ways, in opposition to linear high-speed temporalities. Play and playfulness presuppose the absence of a utilitarian perception of time, the insistence that time should be used carefully and inevitably lead to durable and solid outcomes. That is why the field of "innovation management" is a concept bordering on being an oxymoron, imposing managerial routines on what is supposed to be creative and eventful (see also Weick and Westley [1999] on organizational learning). The concept of reciprocity institutes its own temporal regime whereby the exchange of gifts and counter-gifts is not instantly finalized, but where a gift given today may be returned in years or decades to come. The very norm of reciprocity defies, in fact, an instrumental enactment of time because the very idea of the gift exchange is not to close the gift cycle as soon as possible, but to maintain ongoing and viable social relations. If one gift is instantly reciprocated by a counter-gift, the very idea of gift exchange will be undermined; the two parts are *supposed* to remain bonded by the idea of gift exchange over time or else the idea itself becomes socially insignificant. Moreover, the idea of squandering and expenditure in the general economy is yet another social procedure that, in many ways, undermines the linear cumulative temporality of late modern society. In fact, if time is money and money is the accumulation of resources *qua* capital, the very squandering of resources is the symbolic destruction of something that is essentially non-destructible – time. The slow and tedious ordeal of accumulating resources in everyday life is shifting to short and wasteful periods during, for example, festivals. The symbolic destruction of time in the form of resources symbolizes here the strength and vitality of a particular social formation. In all these cases, the Puritan view of temporality as the slow but ceaseless accumulation of resources is abandoned for other conceptualizations of time, e.g. those enacted in playful existence (essentially being devoid of an experience of time), in reciprocity (where time is deferred, pushed into the future for the benefit of viable and solid social relations), and in squandering (where time is turned into something to be wasted, something to be treated with little reverence). In these three situations, time is no longer perceived as mechanical, equidistant clock-time, but as *duration*, as "enlived" human time – time as it is experienced by the Nuer. Human existence is not of necessity aligned with engineered time; rather, perceived time, duration, is never linear and always intimately connected with past experiences and anticipated futures. In comparison, mechanical clock-time is a dry and lifeless abstraction, an endless series of temporal points moving us forward into a future that holds little more potential than yet another short and transient present moment silently slipping away as the next moment emerges. This overtly abstract mechanical clock-time cannot serve as a basis for human existence, it needs to be complemented by the lived and embodied time of the duration.

Perhaps the single most important reason why the principles of playfulness, reciprocity, and squandering are ignored and overlooked by proponents of instrumental rationalities is precisely this disregard for the mechanical clock-time, so close to the heart of instrumental rationality and its Puritan view of time as something that must not be wasted. Mechanical clock-time has many merits in terms of coordinating many complex social undertakings and individual pursuits, but it is not what makes us human. Instead, the sense of meaningful existence is derived on the basis of duration whereby the past, the present, and the future are intimately connected in ways that provide us with an understanding of human existence in full. As a consequence, the very issue of time and temporality is of great importance when theorizing economic agency. Agency is always executed under the influence of temporal regimes, in most cases combining mechanical clock-time and duration; therefore, the individual and collective enactment of time in many ways influences agency. This thesis can be formulated using two propositions:

(1) In the knowledge-based and aesthetically-driven economy, time and capital no longer have a linear relationship as natural resources – raw materials – they are no longer transformed into commodities; instead, know-how and aesthetic and symbolic resources serve as the basis of economic production.

(2) In what has been called the attention economy, the economy of the abundance of information, managements of organizations and firms are no longer effectively structured using the basis of temporality as the principal parameter; non-linearity displaces the linearity of traditional industrial production.

In fact, playfulness, reciprocity, and squandering are all part of the duration of the organization members. Activities in these spheres of human activity are not overtly concerned with time, but with experiences that transcend mechanical clock-time, that upend the sequential temporality of everyday existence.

Implications for organization studies

Organizational temporalities are not only a practical concern, but also bound up with specific epistemologies inasmuch as enacted temporalities lead to various forms of agency and situational rationalities. Here, we can differentiate between three different temporalities that affect organizational activities: The *circular time* of the premodern period, dominated by the seasonal shifts regulating agrarian production; the *linear time* of modernity, the influence of mechanical clock-time in coordinating work activities based on the principle of a wide-reaching division of labor; the *pluritemporalism* of the late modern, contemporary period, less concerned with agrarian production or manufacturing and more with making knowledge, innovativeness, and creativity the key resources propelling the economy. Thus, pluritemporalism implies that there are a variety of temporalities that co-exist at any given point in time. The temporality of play is non-linear or circular, while the temporality of reciprocity is linear but disruptive, "spatially extended" and stretched out. The temporality of squandering and waste relies on the destruction of time, a sort of inversed reciprocity; while reciprocity values time as the basis for social relations, the squandering of time suggests an alternative view, i.e. that time wasted also affirms society in its own idiosyncratic way. In late modernity, pluritemporalism needs to be combined or aligned in various ways; circular, linear, and non-linear time horizons are brought together in activities that are innovative and creative. Expressed another way, different temporal horizons are enacted within specific regimes of situational rationality and agencies. Townley (2008: 132) defined situational rationality as "the temporally and spatially located sequential and interactional rationality of daily life", a form of practical reason embodied in rules, norms, and institutions. In addition, Hanlon et al. (2010: 166–167) speak of a "substantive rationality" in cases where practical reason are executed within a field of practice. Third, Emirbayer (1997: 294) suggests that "agency is both path-dependent and situationally embedded", i.e., rooted in practical endeavours and interests. In all three cases, the agency of the actors is embedded in situational rationalities, in concerns about practical effects and outcomes. Such agency is not bound up exclusively with either circulatory or linear temporalities, but accommodates, on the contrary, a variety of temporal horizons. In the late modern period, the linear temporality of mechanical clock-time is interrupted due to creative and innovative pursuits being organized, of necessity operating within other temporal horizons.

Wajcman and Rose (2011) demonstrate, for instance, that knowledge workers using new media such as email, voicemail, mobile phones, and so forth, do not regard incoming messages as "disruptive". In the new regime of work, where communication is technologically-mediated rather than face-to-face, what Wajcman and Rose (2011: 956) call "communication episodes" are relatively short, on average lasting five minutes or less. Being able to handle incoming information and communication is a skill learned by knowledge workers:

> [T]he knowledge-workers in our study regard mediated interactions as an essential part of their job roles. These interactions form the basis on which workers are kept up-to-date with tasks and informed of new developments. In this way, they are crucial to the operations of work and can be viewed as positive. (Wajcman and Rose, 2011: 956)

In addition, qualified knowledge workers learn what incoming messages and information to pay attention to and what to deal with later; in doing so, they operate within a pluritemporal horizon. That is, the new media are capable of accommodating and storing time for later use:

> Workers need to negotiate this obstinate presence of information about the potentially multiple mediated interactions that call for their attention. Decisions about the sequential ordering of tasks take into account the perceived urgency of these alerts and the ease with which messages can be stored and later retrieved in the different technologies. (Wajcman and Rose, 2011: 957)

Operating in a technologically-mediated environment, temporality is bound up with the media in use; some communication demands an immediate response while other information may be useful at a later stage. New media thus enable a regime of pluritemporalism. The know-how, skills, and competencies executed in innovative and creative work are not structured in accordance with strictly linear sequences but based on non-sequential modes of thinking.

Stafford (2009: 282), discussing how the limbic system of the brain reinforces "certain perceptual and cognitive constants of reality", making common sense vision a form of ready-made visuality that is incapable of offering new perspectives – a form of lazy and complacent vision comprehending only that which has been seen already – suggests that true creativity lies precisely in the ability to escape such "inherited" modes of seeing. "Creativity may well lie in escaping, not giving in to, our autopoetic machinery and focusing carefully on the world", writes Stafford (2009: 289). Similarly, knowledge-intensive and creative work needs to escape a strict linear temporality, a bureaucratization of time, in order to construct a pluritemporal sphere not fully submitting to mechanical clock-time. Agencies embedded in situational rationalities cope with what is at hand, while at the same time, those activities need to be aligned with overarching organizational and societal activities and objectives. This means that a pluritemporal domain of practice is constructed, one which is capable of aligning alternative time horizons. Such pluritemporal domains of practice are characterized by heterogeneity and coordination. While pre-modern working time was coordinated by the circular recurrence of daily, monthly, or seasonal activities, the modern factory or administrative unit has been structured on the basis of carefully-designed sequences of work and rest. During the contemporary period, it is the working subject him- or herself, monitoring time and making his/her own evaluations of when "enough is enough" and "it's time to go home". As has been demonstrated by, for instance, Maravelias (2003), such an absence of shared and collective temporal structures commonly leads to *more*, not less, hours being worked. The *Selbst-*

zwang of the professional worker, his/her self-discipline and internalized work ethos, easily expand the work-day beyond the hours he/she receives pay for.

While the concept of attention has been highlighted as a resource in short supply in contemporary society, this resource is intimately bound up with two conditions: cognitive limitations and temporal limitations. In fact, the concept of attention is the ratio between these two parameters; limited cognitive limitations but an excess of time entail limited capacities for attention; adequate cognitive limitations but limited time entail equally limited capacities for attention. In many cases, both these conditions prevail. In the regime of endemic time famine (Perlow, 1999), attention will inevitably be in short supply. Today, in the second decade of the new millennium, there is little evidence that the capacity for attention will increase since time will increasingly become a social resource in short supply and since there is a continuous production of new commodities, services, ideas, concepts, and so forth, all competing for our attention. We are living in the attention economy, which is characterized by a surplus of resources.

Ecce Homo Faber: To see the creating human

When it comes to the concept of economic agency and situational rationalities underlying all innovation, the utility-maximizing and calculating *homo oeconomicus* model of agency serves its function in many fields and domains of society, but it remains a poor model for explaining how innovative, creative, and thoughtful ideas come to life. Being bound up with an ethos of self-interest, a strictly linear time perspective, and a calculative and instrumental mind-set, *homo oeconomicus* as an analytical model, an ideal-type, is conspicuously poorly equipped for recognizing elements such as playfulness, reciprocity and squandering in human societies. Playfulness defies calculative and instrumental thinking (even though some games of the *alea* category draw on estimated risk-taking): reciprocity assumes the presence of others in any social setting, largely in opposition to the autonomous position of the *homo oeconomicus*; squandering, finally, is a violation of the utility-maximizing and Puritan worldview that the economic man rests on. In other words, *homo oeconomicus* serves a role when it comes to understanding how human actors behave under determinate conditions, in social situations that are determinate, as in the text-book examples of finance theory whereby actors can make informed choices between risk versus high-yield alternatives. However, just as life does not strictly emerge from a specific combination of amino-acids, but from processes that are not yet widely understood, innovation and creativity do not emerge *ex nihilo*; in the barren landscape of the *homo oeconomicus*, every single act is guided by the intention to maximize personal gain, but such a view does not generate any substantial contributions. Only by recognizing human capacities for playfulness, reciprocity, and squandering – the release of energy, thinking, resources, etc., that are *excessive* to the social system – will new thinking be able to emerge. Any social system absorbing excessive resources in the form of individual gains will gradually lose its capacity to create new ideas and new thinking. The idea of *homo oeconomicus*, the administrative, calculative, short-sighted, and self-interested register of economic agency, needs to be contrasted with and complemented by the human faculties more concerned with playing, sharing, giving away, and short-circuiting the concern for personal gain; in short, a social being who collaborates with others to create the new. It may be that greed is good in directing individual preferences in predictable manners, but one of the shortcomings of greed is that it is too concerned with keeping what may be more effectively put to use when shared. Thus, any theory of economic agency and situational rationality must not only address the personal desires and objectives of human actors (as the *homo oeconomicus* model effectively does), but also the collective interests as regards producing new thinking. Societies that are unable to invest resources in collective endeavors will gradually deteriorate; advanced human societies did not develop until there was a surplus of food that freed up time for more intellectual pursuits (Leroi-Gourhan, 1993), leading to technological innovations that further advanced these human societies and the human species *per se*. Once all resources have been consumed (either immediately or kept as personal property), little will be left for the advancement of new thinking. Economic agency and situational rationalities must not be "undersocialized" (Granovetter, 1985), but must recognize that any social formation needs to be supported by both calculative thinking and procedures and practices for expanding the domain of joint work aimed at creating innovations and creative solutions to perceived problems.

This suggests that organization studies should increasingly examine the role of, for example, playfulness, reciprocity, and squandering in organizational activities. Under what conditions do organizational members act, either legitimately or not, outside of strictly instrumental rationalities? What are the manifest and latent roles of rituals and ceremonies in organizations and professional fields? What forms of waste and expenditure are sanctioned within organizations and industries? The first step toward a research agenda that recognizes such procedures is to de-familiarize the *homo oeconomicus* model of economic agency as a situated and highly-idiosyncratic analytical model developed within neoclassical economic theory. In that perspective, it will be little more than a heuristic designed to advance certain theoretical frameworks and assumptions. As Canguilhem (2009) suggests, since empirical evidence neither constitutes, nor falsifies, theoretical frameworks, being embedded instead in deep-seated beliefs and assumptions, it would be a fallacy to claim that the *homo oeconomicus* model is inadequate or even "false" since it is precisely what is articulated within a specific analytical framework. What limits truth in then no longer what is false, as René Thom points out (cited in Virilio, 2002: 17), but the *insignificant*. An adequate critique of the *homo oeconomicus* model is, then, that it is insignificant when explaining novelty and change, two of the principal drivers of contemporary competitive capitalism. Therefore, rather than seeking to play the role of the iconoclast, overturning incumbent theoretical frameworks and sifting out untruths, the organization theorist should spend more time seeking to articulate theoretical frameworks capable of explaining novelty and change, the explorative practices of organizations, rather than engaging in too much analysis of the exploitative practices, the various calculative and instrumental rationalities put to use in managerial practice. Such a shift in perspective will demand new analytical frameworks and categories.

Concluding remarks

In the field of organization theory and management studies, the question of novelty and creation remains a key concern. While *administration* is a term used for optimizing a specific arrangement of organizational activities, practices and processes, *innovation* denotes the very creation of new ideas and commodities that advance new opportunities and positions for the focal firm. Within a particular administrative regime, it is possible to optimize and fine-tune various managerial practices monitoring the effective use of resources in day-to-day work. In this setting, calculative modes of thinking are applicable in terms of prescribing certain activities, choices, and preferences. Here, the *homo oeconomicus* is the ideal-typical economic actor, capable of rationally calculating the outcome of various scenarios and choosing between comparable choices. When it comes to the question of novelty and creation, *homo oeconomicus* becomes less useful as an analytical model simply because the calculative mode of thinking no longer applies when uncertainty intervenes in decision-making. *Homo oeconomicus* is equipped with a set of rational principles that defy risk-taking when the risk is genuine uncertainty rather than calculable expectancy values; consequently, this analytical model is no longer relevant.

While sociologists like Pierre Bourdieu reject the *homo oeconomicus* model altogether on the basis of the downgrading or elimination of social and cultural concerns when theorizing human agency, one might argue that this analytical model, in fact, has its merits in fields characterized by calculative and computational possibilities. For instance, in the finance industry, financial traders largely act in accordance with the prescribed *homo oeconomicus* model, reports Abolafia (2001). The shortcomings of *homo oeconomicus* become salient when the question of how to create novelty and innovation is addressed, that is, when we shift our perspective from *what is* to *what may become*. Similar to life, undergoing the constant process of changing and becoming, a capitalist regime of accumulation is, according to Marx' analysis, undergoing a constant process of overturning itself in the reproduction of capital. Joseph Schumpeter's (1942) concept of creative destruction is thoroughly indebted to Marx' analysis of capitalism as an economic regime wherein, eventually, "all that is solid melts into thin air". While Schumpeter, a liberal and a representative of the Austrian School of Economics, did not share Marx' view of capitalism as an economic system exploiting the working class and therefore predestined to lead to its own collapse as class conflicts escalate in the face of the unequal distribution of economic resources, he still shared Marx' basic analysis of capitalism as undergoing a constant process of change. In Marx' historical perspective, the bourgeoisie is the only successful revolutionary class in displacing the aristocracy, with its emphasis on respect for traditions and a conservative social outlook that included a restless involvement with creating new products and business opportunities. The aristocracy rested on a feudal and agrarian economy, with limited possibilities for social movement; in the new era of increasing international trade and, eventually, industrialization, the old elite found themselves overtaken by the venturesome bourgeoisie. The capacity to take risks and recognize the new possibilities of the modern period characterized the bourgeoisie mindset and, while the aristocracy assumed their privileges would last forever since they had been part of the social order for centuries, the bourgeoisie did not rest on their laurels. For Schumpeter, as for Marx, it is the bourgeoisie who are the capitalist class *par excellence* in terms of valuing, as well as benefitting from, the creation of new businesses, commodities, and services. However, novelty and creation are the outcome of the expenditure of energy in some form; new ideas, products, and social arrangements demand critical reflection and new thinking.

The ability to transcend what is immediately present demands foresight and vision. Some individuals may embody such capacities but most innovative thinking and new practices are the outcome of joint collaborations. While, for instance, the entrepreneurship literature is concerned with the identification of entrepreneurial skills and qualities (Audretch, 2007; Zott and Huy, 2007; Aldrich, 2005; Acs and Audretch, 2003; Thornton, 1999), enacting the entrepreneur as a figure equipped with the capacity for such foresight and vision, the entrepreneur is relatively rarely an autonomous actor who operates in a social void; on the contrary, he/she is located in a dense social framework wherein he/she is capable of orchestrating complex social interactions in order to accomplish desirable goals (see, for example, Akrich, Callon and Latour, 2002; Hargadon and Douglas, 2001). Entrepreneurial activities, as well as other forms of creation, demand that ideas and thoughts are shared, passed on, discussed, and critically evaluated. Novelty and creation are thus the outcome of a playful attitude whereby thoughts are shared and many ideas are dropped at an early stage. Little new thinking derives from a calculative mode of thinking that operates on the basis of known alternatives; new thinking is, on the contrary, collective, adventurous, and reflective. Similar to the distinction between life and the living, *that-by-which-the living-is-living* and the manifestations of life, new ideas always operate prior to their modalities, their stabilization into social arrangements, commodities, and services. Life *qua* ontological principle is beyond and prior to its manifestations in the living; similarly, ideas are beyond and prior to any new commodity. These ideas are embedded in the jointly-enacted situational rationalities and capacities for economic agency. The principles of playfulness, reciprocity, and squandering underlie such jointly-enacted situational rationalities and capacities for economic agency. Only by maintaining a playful and creative attitude, sharing thoughts and ideas, and by wasting surplus ideas, will economic agents be able to articulate new thinking and new commodities and services. Novelty and creation are not the outcome of optimizing resources at work, or maximizing output, being instead produced outside of such utilitarian frameworks; creation is the effect of serendipity, chance, unpredicted events, idiosyncratic conditions foreseen by no one. The beauty of both life and human accomplishment lies precisely in this unpredictability and in the conditions of emergence, the genuine uncertainty and the surprises facing, in Louis Pasteur's phrasing, even the most "prepared mind". Both the sciences and the world of commerce and business offer a series of narratives regarding unexpected findings and successes. Who would have been able, in the year 2000, to predict the growth in Google's economic value less than a decade hence? Who would have been able make an adequate assessment of the social significance of social media sites such as Facebook a few years ago? As both Marx and, eventually, Schumpeter suggested, the capitalist regime of economic accumulation rests on its remarkable capacity to both appropriate new ideas *and* discard obsolete ones. Creation always casts a shadow of destruction. In analogy with life, capitalism does not suffer from nostalgia, nor is it overtly respectful of past accomplishments, rather it moves on to face new opportunities and challenges. However, just as life-as-principle precedes life-as-manifestation, the commodities and social arrangements of the capitalist regime of accumulation are preceded by new ideas and modes of thinking. Such new ideas and modes of thinking are partly dependent on human beings' capacities for playfulness, reciprocity, and squandering; new thinking is an expenditure of energy and such expenditure is never capable of promising anything but new thinking. The creative destruction of capitalism is thus propelled, not by the administrative routines and calculative practices that constitute its institutional arrangement and infrastructure, but by thinking that operates in parallel with, or even outside of, such calculative worldviews. At the heart of capitalist accumulation lies its antithesis, the expenditure of energy in various forms.

Seen in this light, situational rationalities and economic agency should preferably include both calculative modes of thinking and the capacity to think in new and innovative terms. The capitalist regime of accumulation demands both skills, administration as well as innovation management, in order to subsist. Traditionally, business school curricula have put a lot of emphasis on the calculative practices, the institutionalized routines for reporting economic performance in accounting; however, since creativity and innovation have garnered increased interest, there is a range of courses and training programs today that seeks to train and educate both students and practicing managers in recognizing the importance of creating new commodities and services. In order to recognize how novelty and creation evolve, alternative images of situational rationalities and economic agency need to be developed. Playing, sharing, and wasting are not, in this perspective, irrational social processes. On the contrary, they are part of the infrastructure of the economy; without expenditure, there will be little future gain.

Bibliography

Abdelal, Rawi, (2007), *Capital rules: The construction of the global finance*, Cambridge: Harvard University Press.

Abolafia, Michael, (2001), *Making markets: Opportunism and restraints on Wall Street*, Cambridge: Harvard University Press.

Abramis, David J., (1990), Play at work: Childish hedonism or adult enthusiasm?, *American Behavioral Scientist*, 33(3): 353–373.

Ackroyd, Stephen, (1996), Organization contra organizations: Professionals and organizational change in the United Kingdom, *Organization Studies*, 17(4): 599–621.

Ackroyd, Stephen and Muzio, Daniel, (2007), The reconstructed professional firm: Explaining change in English legal practices, *Organization Studies*, 28(5): 729–747.

Acs, Zoltan J. and Audretch, David B., eds., (2003*)*, *Handbook of entrepreneurship research: An interdisciplinary survey and introduction*, Boston, Dordrecht & London: Kluwer.

Adam, Barbara, (2006), Time, *Theory, Culture and Society*, 23(2–3): 119–138.

Adorno, Theodor W., (1991), *The culture industry: Selected essays on mass culture*, London: Routledge.

Adorno, Theodor W., (2000), *Introduction to sociology*, Cambridge: Polity Press.

Agamben, Georgio, (2009), *What is an apparatus? and other essays*, trans. by David Kishik and Stefan Pedatella, Stanford: Stanford University Press.

Akrich, Madeleine, Callon, Michel and Latour, Bruno, (2002), The key success in innovation part I: the art of interessement, *International Journal of Innovation Management*, 6(2): 187–206.

Alac, Morana, (2008), Working with brain scans: Digital images and gestural interaction in FMRI laboratory, *Social Studies of Science*, 38(4): 483–508.

Alchian, A. and Demsetz, H., (1972), Production, information costs and economic organization, *Amercian Economic Review*, 62(5): 777–795.

Aldrich, Howard E., (2005), Entrepreneurship, in Smelser, Neil J. and Swedberg, Richard, (2005), *The handbook of economic sociology*, Princeton & London: Princeton University Press, pp. 451–477.

Almeling, Renee, (2007), Selling genes, selling gender; Egg agencies, sperm banks, and the medical market in genetic material, *American Scoiological Review*, 73(3): 319–340.

Althusser, Louis, (1984), Ideology and ideological state apparatuses, in Althusser, Louis, (1984), *Essays on ideology*, London & New York: Verso.

Amable, B., (2011), Morals and politics in the ideology of neo-liberalism, *Socio-Economic Review*, 9(1): 3–30.

Anderson, Joseph V., (1994), Creativity and play: A systematic approach to managing innovation, *Business Horizons*, 37(2): 80–85.

Anderson-Gough, Fiona, Grey, Christopher and Robson, Keith, (1998), *Making up accountants: The organizational and professional socialization of trainee chartered accountants*, Aldershot: Ashgate.

Ansell Pearson, Keith, (1999), *Germinal life: The difference and repetition of Deleuze*, London & New York: Routledge.

Argyris, Chris and Schön, Donald A., (1978), *Organizational Learning: A Theory of Action Perspective*, Reading, MA: Addison-Wesley.

Arraghi, Giovanni, (2010), *The long twentieth century*, 2nd ed., London & New York: Verso.

Aspers, Patrik, (2009), Knowledge and valuation in markets, *Theory and Society*, 38: 111–131.

Astley, W. Graham, (1985), Administrative science as socially constructed truth, *Administrative Science Quarterly*, 30: 497–513.

Audretch, David B., (2007), (2006), *The entrepreneurial society*, Oxford: Oxford University Press.

Baba, Yasunori and Walsh, John P., (2010), Embeddedness, social epistemology and breakthrough innovation: the case of the development of statins, *Research Policy*, 39: 511–522.

Bakhtin, Michail, (1968), *Rabelais and his world*, Bloomington: Indiana University Press.

Bakhtin, Michail, (1981), *The dialogical imagination: Four essays*, trans. by Caryl Emerson and Michael Holquist, Austin: University of Texas Press.

Barad, Karen, (2011), Erasers and erasures: Pinch's unfortunate 'uncertainty principle,' *Social Studies of Science*, 41(3): 443–454.

Barley, Stephen R. and Kunda, Gideon, (2006), Contracting: A new form of professional practice, *Acdemy of Management Perspectives*, 20(1): 45–66.

Barley, Stephen R., Meyerson, Debra E. and Grodal, Stine, (2011), E-mail as a source and symbol of stress, *Organization Science*, 22(4): 887–906.

Barrett, Peter and Sexton, Martin, (2006), Innovation in small, projct-based construction firms, *British Journal of Management*, 17: 331–346.

Barthes, Roland, (1981), *Camera Lucida*, trans. by Richard Howard, London: Vintage.

Barry, Andrew, (2005), Pharmaceutical matters: The invention of informed materials, *Theory, Culture & Society*, 22(1): 51–69.

Barry, Andrew and Thrift, Nigel, (2007), Gabriel Tarde, imitation, invention and economy, *Economy and Society*, 36(4): 509–525.

Bartky, Ian R., (1989), The adaptation of standard time, *Technology and Cultre*, 30: 25–56.

Bataille, Georges, (1988a), *Inner Experiences*, Albany: State University of New York Press.

Bataille, Georges, (1988b), *The accursed share: An essay on general economy*, New York; Zone Books.

Bataille, Georges, (1991), *The Accursed Share, Vol. 2: History of Eroticism*, Zone Books, New York.

Bataille, Georges, (2003), *The unfinished system of nonknowledge*, trans. by Michelle and Stuart Kendall, Minneapolis: The University of Minnesota Press.

Baudrillard, Jean, (1998a), *The consumer society: Myths & Structures*, Thousand Oaks, London & New Delhi: Sage.

Baudrillard, Jean, (1998b), When Bataille attacked the metaphysical principles of economy, in Bottig, Fred and Wilson, Scott, eds., (1998), *Bataille: A critical reader*, Oxford: Blackwell.

Bauman, Zygmunt, (2005), *Liquid life*, Cambridge: Polity Press.

Bechky, Beth, (2006), Gaffers, gofers, and grips: Role-based coordination in temporary organizations, *Organization Science*, 17: 3–21.

Becker, Gary S., (1992), The Economic Way of Looking at Life, Nobel Prize Lecture, URL http://nobelprize.org/nobel_prizes/economics/laureates/1992/becker-lecture.pdf (accessed March 11, 2010).

Becker, Howard S., Geer, Blanchie, Highes, Everett C. and Strauss, Anselm L., (1961), *Boys in white: Student culture in medical school*, Chicago: The University of Chicago Press.

Beckert, Jens, (2002), *Beyond the market: The social foundation of economic efficiency*, trans. by Barbara Hershaw, Princeton: Princeton University Press.

Beckert, Jens, (2009), The social order of markets, *Theory and Society*, 38: 245–269.

Beckert, Jens, (2011), The transcending power of goods: Imaginative values in the economy, in Beckert, Jens and Aspers, Patrik, eds., (2011), *The worth of goods. Valuation and pricing in the economy*, Princeton: Princeton University Press, pp. 106–128.

Beller, Jonathan, (2006), *The Cinematic Mode of Production: Attention Economy and the Society of the Spectacle*, Duke Hanover: Dartmouth College Press.

Benjamin, Walter, (1999), *The arcades project*, trans. by Howard Eiland and Kevin McLaughlin, Cambridge: The Belknap Press.

Benner, Patricia, (1984), *From novice to expert: Excellence and power in clinical nursing practice*, San Francisco:Addison-Wesley.

Bennington, Geoffrey, (1995), Introduction to economics I: Because the world is round, in Gill, Carolyn Bailey, (1995), Bataille: Writing the sacred, London & New York: Routledge.

Bennis, Warren G. and O'Toole, James, (2005), How business schools lost their way, *Harvard Business Review*, 83(5): 33–53.

Berchuk, Lucian and Grinstein, Yaniv, (2005), the growth of executive pay, Oxford Review of Economic Policy, 21(2): 283–303.

Bergquist, Magnus and Ljungberg, Jan, (2001), The power of gifts: Organzing social relationships in open source communities, *Information Systems Journal*, 11: 305–320.

Bergson, Henri, (1999), *Laughter*, Copenhagen & Los Angeles: Green Integer Books.

Berle, Adolf A. and Means, Gardiner C., (1934/1991), *The Modern Corporation & Private Property*, New Brunswick. Transaction Publishers.

Best, Steven and Kellner, Douglas, (1991), Postmodern theory: Critical Interrogations, London: Macmillan.

Bijker, Wiebe E., (1995), *Of bicycles, bakelites, and bulbs: Toward a theory of sociotechnical change*, Cambridge & London: The MIT Press.

Bijsterveld, Karin, (2008), *Mechanical sound: technology, culture and public problems of noise in the twentieth century*, Cambridge & London: The MIT Press.

Bijsterveld, Karin and Schulp, Marten, (2004), Breaking into a world of perfection: Innovation in today's classical musical instruments, *Social Studies of Science*, 34(5): 649–674.

Billig, Michael, (2005), *Laughter and ridicule: Towards s social critique of humour*, London, Thousand Oaks & New Delhi: Sage.

Bird-David, Nurit and Darr, Asaf, (2009), Commodity, gift and mass-gift: On gift-commodity hybrids in advanced mass consumption cultures, *Economy and Society*, 38(2): 304–325.

Black, Willam K., (2005), *The best way to rob a bank is to own one: How corporate executives and politicians looted the S&L industry*, Austin: The University of Texas Press.

Blau, Judith R. and McKinley, William, (1979), Ideas, complexity and innovation, *Administrative Science Quarterly*, 24: 200–219.

Blau, Peter M., (1964), *Power and Exchange in Social Life*, New Brunswick: Transaction Books.

Blee, Kathleen M., and Creasap, Kimberly, (2010), Conservative and right-wing movements, *Annual Review of Sociology*, 36: 269–296.

Blum, Virginia L., (2005), *Flesh wounds: the culture of cosmetic surgery*, Berkeley: The University of California Press.

Blumer, Herbert, ([1969], 2007), Fashion: From class differentiation to collective selection, in Barnard, Malcolm, ed., (2007), *Fashion theory: A reader*, London & New York: Routledge, pp. 232–246.

Boczkowski, Pablo J., (2004), *Digitizing the news: Innovations in on-line newspapers*, Cambridge & London: The MIT Press.

Boland, Richard J., Lyytinen, Kalle and Yoo, Youngjin, (2007), Wakes of innovation in project networks: The case of digital 3-D representation in architecture, engineering, and construction, *Organization Science*, 18(4): 631–647.

Bolter, David Jay, (1996), Virtual reality and the redefinition of self, in Strate, Lance, Jacobson, Ronald and Gibson, Stephanie, B., eds., (1996), *Communication and cyberspace: Social interaction in an electronic environment*, Cresskill, NJ: Hampton Press, pp. 105–119.

Bordo, Susan, (2009), Twenty years in the twilight zone, in Heyes, Cressida J. and Jones, Meredith, eds., (2009), *Cosmetic surgery: A feminist primer*, Farnham & Burlingston: Ashgate, pp. 21–33.

Borgerson, Janet and Rehn, Alf, (2004), General economy and productive dualisms, *Gender, Work and Organization*, 11(4): 455–474.

Borges, Jorge Luis, (1999), *Selected non-fiction*, Weinberger, Eliot, ed., London: Penguin.

Boudon, Raymond, (2003), Beyond rational choice theory, *Annual Review of Sociology*, 29: 1–21.

Bourdieu, Pierre, (1993), *The field of cultural production: Essays on art and literature*, edied by Randall Johnson, Cambridge: Polity Press.

Bourdieu, Pierre, (2000), *Pascalian meditations*, Cambridge: Polity Press.

Bourdieu, Pierre, (2005), *The economic structures of society*, Cambridge: Polity Press.

Bower, Frances E., (2002), Organizational slack and corporate greening: Broadening the debate, *British Journal of Management*, 6(1): 29–39.

Bresnen, Mike, Edelman, Linda, Newell, Sue, Scarbrough, Harry and Swan, Jacky, (2005), Exploring social capital in the construction firm, *Building Research and Information*, 33(3): 235–244.

Briggs, Asa and Burke, Peter, (2009), *A social history of the media: From Gutenberg to the Internet*, Cambridge and Malden: Polity.

Britan, Gerald M., (1981), *Bureaucracy and innovation: An ethnography of policy change*, Sage: Beverly Hills.

Brockman, Jeffrey, Chang, Saeyoung and Rennie, Craig, (2007), CEO cash and stock-based compensation changes, layoff decisions, and shareholder value, *Financial Review*, 42: 99–119.

Brody, Howard, (2007), *Hooked: Ethics, the medical profession, and the pharmaceutical industry*, Lanham: Rowman & Littlefield.

Brooks, Abigail, (2004), 'Under the knife and proud of it': An analysis of the normalization of cosmetic surgery, *Critical Sociology*, 30(2): 207–239.

Brown, Andrew D., Kornberger, Martin, Clegg, Stewart R. and Carter, Chris, (2010), 'Invisible walls' and 'silent hierarchies': A case study of power relations in an architecture firm, *Human Relations*, 63(4): 525–549.

Brown, W., (2006), American nightmare: Neoliberalism, neoconservatism, and de-democratization, *Political Theory*, 34(6): 690–714.

Brunsson, Nils, (1982), The irrationality of action and action rationality: Decisions, ideologies and organizational action, *Journal of Management Studies*, 19(1): 29–44.

Brunsson, Nils, (1985), *The Irrational Organization: Irrationality as a Basis for Organizational Action and Change*, New York: Wiley.

Brunsson, Nils, (2006), *Mechanisms of hope: Maintaining the dream of the rational organization*, Copenhagen: Copenhagen Business School Press; Malmö: Liber; Oslo: Universitetsforlaget.

Brynjolfsson, Erik and Saunders, Adam, (2010), *Wired for innovation: How information technology is reshaping the economy*, Cambridge & London: The MIT Press.

Buckhardt, Jacob, ([1860], 1954), *The civilization of the renaissance in Italy*, trans. by S.G.C. Middlemore, New York: The Modern Library.

Burns, T. and Stalker, G.M., (1961), *The Management of Innovation*, Travistock Publications, London.

Burt, Ronald S., (1997), The contingent value of social capital, *Administrative Science Quarterly*, 42: 339–365.

Butler, Judith, (2004), *Undoing gender*, New York & London: Routledge.

Bynum, W.E., (1994), *Science and the practice of medicine in the Nineteenth century*, Cambridge: Cambridge University Press.

Caillois, Roger, (1958/2001a), *Man, play and games*, trans. by Meyer Barash, Urbana & Chicago: University of Illinois Press.

Caillois, Roger, (1961/2001b), *Man and the sacred*, trans. by Meyer Barash, Urbana & Chicago: The University of Illinois Press.

Callinicos, Alex, (2010), *Bonfire of illusions: The twin crises of the liberal world*, Cambridge: Polity.

Callon, Michel, (1980), The state and technical innovation: A case study of the electrical vehicle in France, *Research Policy*, 9: 358–376.

Callon, Michel, (2007), What does it mean to say that economics is performative?, in Mackenzie, Donald A., Muniesa, Fabio and Siu, Lucia., eds., (2007), *Do economists make markets? On the performativity of economics*, Princeton & Oxford: Princeton University Press, pp. 311–357.

Callon, Michel, (2008), Economic markets and the rise of interactive agencement: From prosthetic agencies to habilitated agencies, in Pinch, Trevor and Swedberg, Richard, eds., (2008), *Living in a material world: Economic sociology meets science and technology studies*, Cambridge & London: The MIT Press, pp. 29–56.

Calvino, Italo, (1996), *Six memos for the next millennium*, London: Verso.

Cambrosio, Alberto, Keating, Peter and Mogoutov, Andrei, (2004), Mapping collaborative work and innovation in bioscience: A computer-assisted analysis of antibody reagent workshops, *Social Studies of Science*, 34(3): 325–364.

Canetti, Elias, (1992), *Crowds and Power*, London: Penguin.

Canguilhem, Georges, (2008), *Knowledge of life*, trans. by Stefano Geroulanos and Daniela Ginsburg, New York: Fordham University Press.

Capaldo, Antonio, (2007), Network structure and innovation: The leveraging of a dual network as a distinctive relational capability, *Strategic Management Journal*, 28: 585–608.

Cardinal, Laura B., (2001), Technological innovation in the pharmaceutical industry: The use of organizational control in managing research and development, *Organization Science*, 12(1): 19–36.

Cardinal, Laura B., Alessandri, Todd M. and Turner, Scott F., (2001), Knowledge codifiability, resources and science-based innovation, *Journal of Knowledge Management*, 5(2): 195–204.

Carrier, James, (1991), Gift commodities and social relations: A Maussian view of exchange, *Sociological Forum*, 6: 119–136.

Carruthers, Bruce and Espeland, Wendy, (1991), Accounting for rationality; Double-entry book-keeping and the rhetoric of economic rationality, *American Journal of Sociology*, 97(1): 31–69.

Casper, Steven and Matraves, Catherine, (2003), Institutional frameworks and innovation in the German and UK Pharmaceutical industry, *Research Policy*, 32: 1865–1879.

Castilla, Emilio J., (2008), Gender, race and meritocracy in organizational careers, *American Journal of Sociology*, 113(6): 1479–1526.

Caves, Richard E., (2000), *Creative industries*, Cambridge & London: Harvard University Press.

Chapkis, Wendy, (1997), *Live sex acts: Women performing erotic labor*, New York & London: Routledge.

Cheal, David, (1988), *The gift economy*, London & New York: Routledge.

Chesbrough, Henry W., (2003), *Open innovation: The new imperative for creating and profiting from technology*, Boston: Harvard Business School Press.

Chorev, Nitsan and Babb, Sarah, (2009), The crisis of neoliberalism and the future of international institutions; A comparison of the IMF and the WTO, *Theory and Society*, 38: 459–484.

Clarke, Simon, (2005), The neoliberal theory of society, in Saad-Filho, Alfredo and Johnston, Deborah, eds., (2005), *Neoliberalism: A critical reader*, London & Ann Arbor: Pluto, pp. 50–59.

Clifford, James, (1988), *The predicament of culture: Twentieth-century ethnography, literature, and art*, Boston: Harvard University Press.

Clifford, James, (1997), *Routes: Travel and translation in the late twentieth century*, Cambridge: Harvard University Press.

Coase, Ronald H., (1937/1991), The nature of the firm, in Williamson, Oliver E. and Winter, Sidney G., eds., (1991), *The nature of the firm: Origin, evolution, and development*, New York & Oxford: Oxford University Press.

Cohen, Margot and Athavaley, Anjali, (2009), A Search for a Surrogate Leads to India, *Wall Street Journal*, Oct. 8, 2009.

Coleman, J.S., (1988), Social capital in the creation of human capital, *American Journal of Sociology*, 94: S95–121.

Collins, Harry and Pinch, Trevor, (2005), *Dr. Golem: How to think about medicine*, Chicago & London: The University of Chicago Press.

Collins, Randall, (1993), Emotional energy as the common denominator of rational choice, *Rationality and Society*, 5: 203–220.

Collinson, David L., (2002), Managing humour, *Journal of Management Studies*, 39(3): 269–288.

Combs, Rod and Hull, Richard, (1990), 'Knowledge management practices' and path-dependency in innovation, *Research Policy*, 27: 237–253.

Comte, Auguste, (1975), *Auguste Comte and positivism: Essential writings*, ed. by Gertrud Lenzer, New York: Harper Torchbooks.

Cooley, Charles Horton, (1902/1964), *Human nature and the social order*, New York. Schocken Books.

Cooper, Cecily D., (2005), Just joking around? Employee humor expressions an ingratiatory behavior, *Academy of Management Review*, 30(4): 765–776.

Cooper, Robert, (2005), *Relationality, Organization Studies*, 26(11): 1689–1710.

Cooren, François, (2006), Organizational world as plenum of agencies, in Cooren, François, Taylor, James R. and Van Every, Elizabeth J., eds., (2006), *Communication as organizing: Empirical and theoretical explorations in the dynamic of text and conversation*, Mahwah & London: Lawrence Erlbaum, pp. 81–100.

Cowan, Robin and Jonard, Nicholas, (2009), Knowledge portfolios and the organization of innovation networks, *Academy of Management Review*, 34(2): 320–342.

Craig, Tim, (1995), Achieving innovation through bureaucracy: Lessons from the Japanese brewing industry, California Management Review, 38(1): 8–36.

Crary, Jonathan, (1999), *Suspensions of perception: Attention, spectacle, and modern culture*, Cambridge & London: The MIT Press.

Critchley, Simon, (2007), Humour as practically enacted theory, or, why critics should tell more jokes, in Westwood, Robert and Rhodes, Carl, eds., (2007), *Humour, work and organization*, London & New York: Routledge, pp. 17–32.

Crossan, Mary M. and Apaydin, Marina, (2010), A multi-dimensional framework of organizational innovation; A systematic review of the literature, *Journal of Management Studies*, 47(6): 1154–1191.

Crotty, James, (2005), The neoliberal paradox: The impact of destructive product market competition and the 'modern' financial markets on nonfinancial corporation performance in the neoliberal era, in Epstein, G., ed., (2005), *The financialization of the world economy*, Northampton: Edward Elgar, pp. 77–110.

Cunliffe, Ann L., Luhman, John T. and Boje, David M., (2004), Narrative temporality: Implications for organizational research, *Organization Studies*, 25(2): 261–286.

Cunliffe, Ann L., (2008), Orientation to social constructionism: Relationality responsive social constructionism and its implications for knowledge and learning, *Management Learning*, 39(2): 123–139.

Currah, Andrew, (2007), Managing creativity: The tension between commodities and gifts in a digital networked environment, *Economy and Society*, 36(3): 467–494.

Cyert, Richard M. and March, James G., (1963), *A Behavioral Theory the Firm*, Englewood Cliffs: Prentice-Hall.

Czarniawska, Barbara, (2009), *Den tysta fabriken*, Malmö: Liber.

Dahlander, Linus and Gann, David M., (2010), How open is innovation?, *Research Policy*, 39: 699–709.

Dalton, Melville, (1959), *Men who manage: Fusion of feeling and theory in administration*, New York: Wiley.

Darr, Asaf, (2003), Gifting practices and interorganizational relations: Constructing obligation networks in the electronics sector, *Sociological Forum*, 18(1): 47–67.

Davies, William and McGoey, Linsey, (2012), Rationalities of ignorance: On financial crisis and the ambivalence of neo-liberal epistemology, *Economy and Society*, 41(1): 64–83.

Davis, Gerald F., (2009), *Managed by the markets: How finance reshaped America*, New York & Oxford: Oxford University Press.

Davis, Kathy, (2009), Revisting feminist debates on cosmetic surgery: Some reflections on suffering, agency, and embodied difference, in Heyes, Cressida J. and Jones, Meredith, eds., (2009), *Cosmetic surgery: A feminist primer*, Farnham & Burlingston: Ashgate, pp. 35–47.

Davis, Mike, (2006), *Planet of slums*, London: Verso.

Davis, Gerald F., (2009b), The rise and fall of finance and the end of the society of organizations, *Academy of Management Perspectives*, 23(3): 27–44.

Davis, Gerald F., Diekmann, Kristine A. and Tinsley, Catherine, (1994), The decline and fall of the conglomerate firm in the 1980s: the deinstitutionalization of an organization form, *American Sociological Review*, 59: 547–570.

Davis, Gerald and Robbins, Gregory, (2005), Nothing but net? Networks and status in corporate governance, in Knorr Cetina, Karin and Preda, Alex, eds., (2005), *The sociology of financial markets*, Oxford & New York. Oxford University Press. pp. 290–311.

Debord, Guy, (1977), *Society of the Spectacle*, Detroit: Black & Red.

Derrida, Jacques, (1981), *Dissemination*, Chicago & New York: The University of Chicago Press.

Derrida, Jacques, (1987), *The truth in painting*, trans. by Geoff, Bennington, and Ian, McLeaod, Chicago & London: The University of Chicago Press.

Derrida, Jacques, (1992), *Given Time: I. Counterfeit Money*, Chicago: The University of Chicago Press.

Derrida, Jacques, (1998), From the restricted to the general economy: A Hegelianism without reserve, in Bottig, Fred and Wilson, Scott, eds., (1998), *Bataille: A critical reader*, Oxford: Blackwell.

Derrida, Jacques and Stiegler, Bernard, (2002), *Echographies of television: Filmed interviews*, trans. by Jennifer Bajorek, Cambridge: Polity.

Dobbin, Frank and Sutton, John R., (1998), The strength of a weak state: The rights revolution and the rise of human resources management divisions, *American Journal of Sociology*, 104(2): 441–476.

Dobbin, Frank and Zorn, Dirk, (2005), Corporate malfeasance and the myth of shareholder value, *Political Power and Social Theory*, 17: 179–198.

Dodgson, Mark, (2000), *The management of technological innovation*, Oxford & New York: Oxford University Press.

Dodgson, Mark, Gann, David and Salter, Ammon, (2005), *Think, play, do: technology, innovation, and organization*, Oxford & New York: Oxford University Press.

Doel, Marcus A., (2009), Misearly thinking/excessful geography: From restricted economy to global financial, crisis, *Environemnt and Planning D: Society and Space*, 27: 1054–1073.

Dore, Ronald, (2008), Financialization of the global economy, *Industrial and Corporate Change*, 17(6): 1097–1112.

Dougherty, Deborah and Bowman, Edward H., (1995), The effects of organizational downsizing on product innovation, *California Management Review*, 37(4): 28–44.

Dougherty, Deborah and Takacs, C. Helen, (2004), Team play: Heedful interrelating as the boundary for innovation, *Long Range Planning*, 37(6): 569–590.

Douglas, Mary, (1966), *Purity and danger: An analysis of concepts of pollution and taboo*, London & Henley: Routledge & Kegan Paul.

Douglas, Mary, (1986), *How institutions think*, London: Routledge & Kegan Paul.

Douglas, Mary and Isherwood, Baron, (1979), *The world of goods: Towards an anthropology of consumption*, London: Allen Lane.

Downer John, (2011), "737-Cabriolet": The limits of knowledge and the sociology of inevitable failure, *American Journal of Sociology*, 117(3): 725–762.

Dreher, George F., (2003), Breaking the glass-ceiling: The effects of sex ratios and work-life programs on female leadership at the top, *Human Relations*, 56(5): 541–562.

Duhem, Pierre, (1996), *Essays in the history and philosophy of science*, trans. and ed. by Roger Ariew and Peter Barker, Indianapolis & Cambridge: Hackett Publishing.

Duménil, Gérard and Lévy, Dominique, (2004), *Capital resurgent: Roots of the neoliberal revolution*, trans. by Derek Jeffer, Cambirdge: Harvard University Press.

Durkheim, Èmile, (1912/1995), *The Elementary Form of Religious Life*, New York: Free Press.

Eco, Umberto, (1983), *The Name of the Rose*, New York: Harcourt.

Eisenstein, Elizabeth, (1983), The *printing revolution in early modern Europe*, Cambridge: Cambridge University Press.

Elias, Norbert, (1978), *The civilizing process*, Oxford: Blackwell.

Elias, Norbert and Dunning, Eric, (1986), *Quest for excitement: Sport and leisure in the civilizing process*, Oxford: Blackwell.

Elsbach, Kimberly D., (2009), Identity affirmation through signature style: A study of toy car designers, *Human Relations*, 62(7): 1041–1072.

Emirbayer, Mustafa, (1997), A manifesto for relational sociology, *American Journal of Sociology*, 103(2): 281–317.

Emirbayer, Mustafa and Mische, Ann, (1998), What is agency?, *American Journal of Sociology*, 103(4): 962–1023.

Entwistle, Joanna, ([2000], 2007), Power dressing and the construction of the career woman, in Barnard, Malcolm, ed., (2007), *Fashion theory: A reader*, London & New York: Routledge, pp. 208–219.

Entwistle, Joanna & Racamora, Agnès, (2006), The field of fashion materialized: A study of London Fashion Week, *Sociology*, 40(4): 735–751.

Epple, Moritz, (2011), Between timelessness and historicality: On the dynamics of the epistemic objects of mathematics, *Isis*, 102: 481–493.

Epstein, Gerald A. and Jayadev, Arjun, (2005), The rise of the rentier income in OECD countries: Financialization, central bank policy and labor solidarity, in Epstein, G., ed., (2005), *The financialization of the world economy*, Northampton: Edward Elgar, pp. 46–74.

Epstein, Stephan R., (1998), Craft guilds, apprenticeship, and technological change in preindustrial Europe, *Journal of Economic History*, 53(4): 684–718.

Erickson, Carolly, (1976), *The medieval vision: Essays in history and perception*, New York.

Ericson, Richard, Barry, Dean and Doyle, Aaron, (2000), The moral hazard of neo-liberalism: Lessons from the private insurance industry, *Economy and Society*, 29(4): 532–558.

Espeland, Wendy Nelson and Stevens, Mitchell L., (1998), Commensuration as a social process, *Annual Review of Sociology*, 24: 13–24.

Espeland, Wendy Nelson and Stevens, Mitchell L., (2008), A sociology of quantification,. *European Journal of* Sociology, 49(3): 401–436.

Fagerberg, Jan, Mowery, David C. and Nelson, Richard R., eds. *The Oxford handbook of innovation*, Oxford & New York: Oxford University Press.

Fama, Eugene F., (1970), Efficent capital markets: A review of theory and empirical work, *Journal of Finance*, 25(2): 383–417.

Fama, Eugene F. and Jensen, Michael, (1983), Separation of ownership and control, *Journal of Law and Economics*, 26(2): 301–325.

Faulkner, Philip and Runde, Jochen, (2009), On the identity of technological objects and user innovation in function, *Academy of Management Review*, 34(3): 442–462.

Febvre, Lucien and Martin, Henri-Jean, ([1958], 1997), *The coming of the book: The impact of printing 1450–1800*, trans. by David Gerard, London & New York: Verso.

Feldman, Steven P., (1989), The broken wheel: the inseparability of autonomy and control in innovation within organizations, *Journal of Management Studies*, 26(2). 83–102.

Feldman, Stephen P., (2004), The culture of objectivity: Quanitification, uncertainty, and the evaluation of risk at NASA, *Human Relations*, 57(6): 691–718.

Ferrary, Michel, (2003), The gift exchange in the social networks of Silicon Valley, *California Management Review*, 45(4): 120–138.

Ferray, Michael and Granovetter, Mark, (2009), The role of venture capital firms in Silicon Valley's complex innovation network, *Economy and Society*, 38(2): 326–359.

Fine, Gary-Alan, (2007), *Authors of the Storm: Meteorologists and the Culture of Prediction*, Chicago & London: The University of Chicago Press.

Fischer, Claude S., (1992), *America calling: A social history of the telephone to 1940*, Berkeley and Los Angeles: University of California Press.

Fischer, Jill, (2009), *Medical research for hire: The political economy of pharmaceutical clinical trials*, New Brunswick: Rutgers University Press.

Fishman, Jennifer R., (2004), Manufacturing desire. The commodification of female sexual dysfunction, *Social Studies of Sicence*, 34(2): 187–218.

Fiss, Peer C. and Zajac, Edward, J., (2004), The diffusion of ideas over contested terrains: The (non)adoption of a shareholder value orientation among German firms, *Administrative Science Quarterly*, 49: 501–534.

Fleck, Ludwik, (1979), *Genesis and development of a scientific fact*, Chicago & London: Chicago University Press.

Fleischer, Anne, (2009), Ambiguity and the equity of rating systems: Unites States brokerage firms, 1995–2000, *Administrative Science Quarterly*, 54(4): 555–574.

Fleming, Peter, (2005), 'The kindergarten cop': paternalism and resistance in a high-commitment workplace, *Journal of Management Studies*, 42(6): 1469–1489.

Fleming, Peter and Sturdy, Andrew, (2011), 'Being yourself' in the electronic sweatshop: new forms of normative control, *Human Relations*, 64(2): 177–200.

Fligstein, Neil, (1987), The interorganizational power struggle: Rise of financial personnel to top leadership in large corporations, *Amercian Sociological Review*, 52(1): 44–58.

Fligstein, Neil, (1990), *The transformation of corporate control*, Cambridge & London: Harvard University Press.

Fligstein, Neil, (2001), *The architecture of markets*, Princeton: Princeton University Press.

Fligstein, Neil and Shin, Taekjin, (2007), Shareholder value and the transformation of the U.S.: economy, 1984–2000. *Sociological Forum*, 22: 399–424.

Florida, Richard, (2002), *The rise of the creative class*, New York. Basic Books.

Floyd, Steven W. and Woolridge, Bill, (1994), Dinosaurs or dynamos? Recognizing middle management's strategic role, *Academy of Management Executive*, 8(4): 47–57.

Flusser, Vilém, (2002), *Writings*, edited by Andreas Ströhl, trans. by Erik Eisel, Minneapolis & London: The University of Minnesota Press.

Foucault, Michel, (2008), *The birth of biopolitics: Lectures at the Collège de France, 1978–1979*, ed. by Michael Senellart, trans. by Graham Burchell, Basingstoke: Palgrave.

Fourcade, Marion, (2009), *Economists and societies: Discipline and profession in the United States, Britain, and France, 1890s to 1990s*, Princeton & London: Princeton University Press.

Fourcade, Marion, (2011), Cents and sensibility: Economic valuation and the nature of 'nature', *American Journal of Sociology*, 116(6): 1721–1777.

Frank, Thomas, (1997), *The conquest of cool: Business culture, counterculture and the rise of hip consumerism*, Chicago & London: The University of Chicago Press.

Frank, Thomas, (2008), Rent-a-Womb Is Where Market Logic Leads, *Wall Street Journal*, Dec. 10, 2008.

Franz, Kathleen, (2005), *Tinkering: Customers invent the early automobile*, Philadelphia: University of Pennsylvania Press.

Freese, Jeremy, (2008), Genetics and the social science explanation of individual outcomes, *American Journal of Sociology*, 114(S): S1–S35.

Friedman, Milton, ([1962], 2002), *Capitalism and freedom*, 4th ed., Chicago & London: The University of Chicago Press.

Froud, Julie, Haslam, Colin, Johal, Sukhdev and Williams, Karel, (2000), Shareholder value and Financialization: Consultancy promises, management moves, *Economy and Society*, 29(1): 80–110.

Furman, Jeffrey L. and MacGarvie, Megan, (2009), Academic collaboration and organizational innovation; the development of research capabilities in the US pharmaceutical industry, 1927–1946, *Industrial and Corporate Change*, 18(5): 929–961.

Galbraith, John Kenneth, (1958), *The Affluent Society*, Houghton Mifflin, Boston.

Gammon, Earl, (2010), Nature as adversary: The rise of the modern economic conception of nature, *Economy and Society*, 39(2): 218–246.

Gann, David M., (2000), *Building innovation: Complex constructs in a changing world*, London: Thomas Telford.

Gantt, Henry L., (1913/1919), *Work, wages, and profits*, 2nd ed., New York: The Engineering Magazine Co.

Garcia-Parpet, Marie-France, (2007), The social construction of a perfect match: The strawberry auction at Fontaines-en-Sologne, in Mackenzie, Donald A., Muniesa, Fabio and Siu, Lucia., eds., (2007), *Do economists make markets? On the performativity of economics*, Princeton & Oxford: Princeton University Press, pp. 20–53.

Garfinkel, Harold, (1967), *Studies in Ethnomethodology*, Prentice-Hall: Englewood Cliffs.

Gassman, O., Reepmeyer, G. and von Zedtwitz, M., (2004), *Leading pharmaceutical innovations: trends and drivers for growth in the pharmaceutical industry*, London: Springer.

Gay, Peter, (1984), *The bourgeois experience: Victoria to Freud: Vol. 1: Education of the senses*, New York & Oxford: Oxford University Press.

Gee, Sophia, (2010), *Making waste: Leftover and the Eighteenth-century imagination*, Princeton & Oxford: Princeton University Press.

Geertz, Clifford, (1978), The bazaar economy: Information and search in peasant marketing, *Amercian Economic Review*, 68(2): 28–32.

Geertz, Clifford, (2000), *Available light: Anthropological reflections on philosophical topics*, Princeton: Princeton University Press.

Geiger, Scott W. and Cashen, Luke H., (2002), A multidimensional examination of slack and its impact innovation, *Journal of Managerial Issues*, 14(1): 68–84.

Gephart, Robert P. Jr., (1996), Postmodernism and the future history of management, *Journal of Management History*, 2(3): 90–96.

Gibson, Christina B. and Gibbs, Jennifer L., (2006), Unpacking the concept of virtuality: The effects of geographic dispersion, electronic dependence, dynamic structure, and national diversity in team innovation, *Administrative Science Quarterly*, 51: 451–495.

Gibson, James J., (1977), The theory of affordances, in Shaw, Robert and Bransford, John, eds., (1977), *Perceiving, knowing, acting: Toward an ecological* psychology, Lawrence Erlbaum, pp. 7667–82.

Gilbreth, Frank B., (1911), *Motion study: A method for increasing the efficiency of the workman*, NewYork: Van Nostrand.

Girard, René, (1977), *Violenace and the sacred*, trans. by Patrick Gregory, Baltimore & London: John Hopkins University Press.

Goffman, Erwin, (1961), *Asylums*, London: Penguin.

Goldstein, Adam, (2012), Revenge of the managers: Labor cost-cutting and the paradoxical resurgence of managerialism in the shareholder value era, 1984 to 2001, *American Sociological Review*, 77(2): 268–294.

Goodwin, Charles, (1995), Seeing in depth, *Social Studies of Science*, 25: 237–274.

Gordon, David, (1996), *Fat and mean: The corporate squeeze of working Americans and the myth of managerial downsizing*, New York: Free Press.

Gotsi, Manto, Andriopoulos, Lewis, Marianne W. and Ingram, Amy E., (2010), managing creaitves: paradoxical appraoches to identity regulation, *Human Relations*, 63(6): 781–805.

Gouldner, Alvin W., (1954a), *Patterns of industrial bureaucracy*, Glencoe: The Free Press.

Gouldner, Alvin W., (1954b), *Wildcat strike*, New York, Evanston & London: Harper Torchbooks.

Gouldner, Alvin W., (1960), The norm of reciprocity; A preliminary statement, *American Sociological Review*, 25(2): 161–178.

Goux, Jean-Joseph, (1998), General economy and postmodern capitalism, in Bottig, Fred and Wilson, Scott, eds., (1998), *Bataille: A critical reader*, Oxford: Blackwell.

Gower, Timothy, (2002), *Mathematics: A very short introduction*, Oxford & New York: Oxford University Press.

Graham, L., (1995), *On the Line at Subaru-Isuzu: The Japanese Model and the American Worker*, ILR press, Ithaca.

Granovetter, Mark, (1985), Economic action and social structure, *American Journal of Sociology*, 91(3): 481–510.

Grant, Adam M. and Berry, James W., (2011), The necessity of others is the mother of invention: Intrinsic and prosocial motivations, perspective taking, and creativity, *Academy of Management Journal*, 54(1): 73–96.

Grasseni, Cristina, (2004), Skilled vision: An apprenticeship in breeding aesthetics, *Social Anthropology*, 12(1): 41–55.

Gray, Jeremy, (2008), *Plato's ghost: The modernist transformation of mathematics*, Princeton, Princeton University Press.

Greenley, Gordon E. and Oktemgil, Mehmet, (1998), A comparison of slack resources in high and low performing British companies, *Journal of Management Studies*, 35(3): 377–398.

Greve, Henrich R. and Taylor, Alva, (2000), Innovations as catalysts for organizational change: Shifts in Organizational cognition and search, *Administrative Science Quarterly*, 45: 54–80.

Griffin, Penny, (2009), *Gendering the World Bank: Neoliberalism and ther gendered foundation of global governance*, New York: Palgrave.

Grimaldi, Rosa, Kenney, Martin, Siegel, Donald S., and Wright, Mike, (2011), 30 years after Bayh–Dole: Reassessing academic entrepreneurship, *Research Policy*, 40: 1045– 1057.

Grint, Keith, (1997), *Fuzzy management: Contemporary ideas and practices at work*, Oxford: Oxford University Press.

Gross, Neil, Medvetz, Thomas and Russeel, Rupert, (2011), The contemporary American conservative movement, *Annual Review of Sociology*, 37: 325–354.

Grosz, Elizabeth, (2004), *The nick of time: Politics, evolution and the untimely*, Durham: Duke University press.

Grosz, Elizabeth, (2011), *Becoming undone: Darwinian reflections on life, politics, and art*, Durham & London: Duke University Press.

Grugulis, Irena, (2002), Nothing serious? Candidates' use of humour in management training, *Human Relations*, 55(4): 387–406.

Guillén, Mauro F., Collins, Randall, England, Paula and Meyer, Marshall, (2002), The revival of economic sociology, in Guillén, Mauro F., Collins, Randall, England, Paula and Meyer, Marshall, eds., (2002), *The new economic sociology: Development in emerging fields*, New York: Russell Sage Foundation, pp. 1–32.

Gurevich, Georges, (1964), *The spectrum of social time*, trans. by Myrtle Korenbaum, Dordrecht: D. Riedel.

Hacking, Ian, (2010), Husserl on the origin of geometry, in Hyder, David and Rheinberger, Hans-Jörg, eds., (2010), *Science and the life world: Essays on Husserl's 'Crisis of European Science'*, Stanford: Stanford University Press, pp. 64–82.

Hallett, Tim and Ventresca, Marc J., (2006), How institutions form: Loose couplings as mechanism in Gouldner's Patterns of Industrial Bureaucracy, *American Behavioral Scientist*, 49: 908–927.

Hanlon, Gerard, Stranglemen, Tim, Goode, Jackie, Luff, Donna, O'Cathlain, Alicia and Greatbatch, David, (2005), Knowing, technology and nursing: The case of NHS direct, *Human Relations*, 58: 147–171.

Hansen, Miriam Bratu, (1995), America, Paris, the Alps: Kracauer (and Benjamin) on cinema and modernity, in Carney, Leo and Schwartz, Vanessa R., eds., (1995), *Cinema and the invention of modern life*, Berkeley: University of California Press, pp. 362–402.

Hargadon, Andrew B. and Douglas, Yellowlees, (2001), When innovations meet institutions; Edison and the design of the electric light, *Administrative Science Quarterly*, 46: 476–501.

Harison, Elad and Koski, Heli, (2010), Applying open innovation in business strategies. Evidence from Finnish software firms, *Research Policy*, 39: 351–359.

Harris, Martin, (2006), Technology, innovation and post-bureaucracy: the case of the British Library, *Journal of Organization Change Management*, 19(1): 80–92.

Harvey, David, (2005), *A brief history of neoliberalism*, New York: Oxford University Press.

Harvey, David, (2010), *The enigma of capital and the crisis of capitalism*, London: Profile Books.

Hatch, Mary Jo, (1997), Irony and the social construction of contradiction in the humour of a management team, *Organization Science*, 8(3): 275–288.

Hayward, Matthew L.A. and Boeker, Warren, (1998), Power and conflict of interest in professional firms: Evidence from investment banking, *Administrative Science Quarterly*, 43: 1–22.

Healy, David, (2006), The new medical oikumene, in Petryna, Adriana, Lakoff, Andrew and Kleinman, Arthur, eds., (2006), *Global Pharmaceuticals: Ethics, Markets, Practices*, Durham & London: Duke University Press, pp. 61–84.

Healy, Kieran, (2004), Altruism as an organizational problem: The case of organ procurement, *American Sociological Review*, 69: 387–404.

Hechter, Michael and Kanazawa, Satishi, (1997), Sociological rational choice theory, *Annual Review of Sociology*, 23: 191–214.

Hegarty, Paul, (2000), *George Bataille: Core cultural theorist*, London, Thousand Oaks & New Delhi: Sage.

Heidegger, Martin, (1961/1987), *Nietzsche, Vol III: The will to power as knowledge and as metaphysics*, trans. by Joan Stambaugh, David Farrell Krell and Frank A. Capuzzi, San Francisco: Harper & Row.

Heidegger, Martin, (1981), "Only God can save us": The *Spiegel* Interview (1966), in Sheehan, Thomas, ed., (1981), *Heidegger: The man and his works*, Chicago: Precedent, pp. 45–75.

Helmreich, Stefan, (2011), What was life? Answers from three limit biology, *Critical Inquiry*, 37(4): 671–696.

Hennion, Antoine, (2002), Music lovers: Taste as performance, *Theory, Culture & Society*. 18(5) 1–22.

Herold, David M., Jayaraman, Narayanan and Narayanaswamy, C.R., (2006), What is the relationship between organizational slack and innovation?, *Journal of Managerial Issues*, 18(3): 372–39.

Hesmondhalgh, David and Baker, Sarah, (2008), Creative work and emptional labour in the television industry, *Theory, Culture and Society*, 25(7–8): 97–118.

Heyes, Cressida J., (2009), Diagnosing culture: Body dysmorphic disorder and cosmetic surgery, *Body & Society*, 15(4): 73–93.

Hines, Ruth D., (1988), Financial accounting: In communicating reality, we construct reality, *Accounting, Organization and Society*, 13: 251–261.

Ho, Karen, (2009), *Liquidated. An ethnography of Wall Street*, Durham & London: Duke University Press.

Hochschild, Arlie Russell, (1983), *The Managed Heart*, Berkeley: University of California Press.

Hochschild, Arlie Russell, (1997), *The time bind: When work becomes home and home becomes work*, New York: Metropolitan Books.

Holquist, Michael, (2003), Dialogism and aesthetics, in Gardiner, Michael E., ed., (2003), *Mikhail Bakhtin*, vol. 1, London, Thousand Oaks & New Delhi: Sage, pp. 367–385.

Holzner, Burkart, (1968), *Reality construction in society*, Cambridge: Schenkman.

Huizinga, Johan, (1949), *Homo Ludens: A study of the play-element in culture*, London: Routledge & Kegan Paul.

Huizingh, Eelko K.R.E., (2011), Open innovation: State of the art and future perspectives, *Technovation*, 31: 2–9.

Håkansson, Lars, (2007), Creating knowledge: The power and logic of articulation, *Industrial and Corporate Change*, 16: 51–88.

Håkansson, Lars, (2010), The firm as an epistemic community: The knowledge-based view revisted, *Industrial and Corporate Change*, 19(6): 1801–1828.

Israel, Paul, (1992), *From machine shop to industrial laboratory: Telegraphy and the changing context of American innovation, 1830–1920*, Baltimore & London: The Johns Hopkins University Press.

James, William, (1890/1950), *Principles of psychology*, vol. 1, New York: Dover.

James, William, (1978), The sentiment of rationality, in James, William, (1978), *Essays in philosophy*, Cambridge & London: Harvard University Press.

Jaques, Elliott, (1951), *The changing culture of a factory*, London: Tavistock Publications.

Jasanoff, Sheila, (2005), The idiom of co-production, in Jasanoff, Sheila, ed., (2005), *States of knowledge: the co-production of science and social order*, London & New York: Routledge, pp. 1–12.

Jassawalla, Avan R. and Sashittal, Hemant C., (2002), Cultures that support product innovation processes, *Academy of Management Executive*, 16(3): 42–54.

Jemielniak, Dariusz, (2009), Time as symbolic currency in knowledge work, *Information & Organization*, 19: 277–293.

Jemielniak, Dariusz, (2012), *The new knowledge workers*, Cheltenham & Northampton; Edward Elgar.

Jensen, M., (1993), The modern industrial revolution, exit, and failure of internal control systems, *Journal of Finance*, 48(3): 831–880.

Jensen, Michael C. and Meckling, William H., (1976), Theory of the firm: Managerial behavior, agency costs and ownership structure, Journal of Financial Economics, 3(4): 305–260.

Jensen, Michael C. and Murphy, Kevin J., (1990), CEO incentives: it's not how much you pay but how, *Harvard Business Review*, 68(3): 138–149.

Johns, Adrian, (1998), *The nature of the book: Print and knowledge in the making*, Chicago & London: The University of Chicago Press.

Jones, Meredith, (2008), *Skintight: An anatomy of cosmetic surgery*, Oxford & New York: Berg.

Jones, Owen, (2011), *Chavs: The demonization of the working class*, London: Verso.

Kahneman, Daniel, (2011), *Thinking, fast and slow*, New York: Farrar, Straus and Giroux.

Kay, Lily E., (2000), *Who wrote the book of life? A history of the genetic code*, Stanford & London: Stanford University Press.

Keller, Evelyn Fox, (1983), *A feeling for the organism: The life and work of Barbara McClintock*, New York & San Francisco: W.H. Freeman.

Kellogg, Katherine C., (2009), Operating room: Relational spaces and microinstitutional change in surgery, *American Journal of Sociology*, 115(3): 657–771.

Kellogg, Katherine C., (2011), Hot lights and cold steel: Cultural and political toolkits for practice change in surgery, *Organization Science*, 22(2): 482–502.

Khurana, Rakesh, (2002), *Searching for a corporate savior: The irrational quest for a charismatic CEO*, Princeton: Princeton University Press.

Khurana, Rakesh, (2007), *From higher aims to hired hands: The social transformation of American business schools and the unfulfilled promise of management as a profession*, Princeton: Princeton University Press.

Kieser, Alfred and Leiner, Lars, (2009), Why the rigour-relevance gap in management research is unbridgeable, *Journal of Management Studies*, 46(3): 516–533.

Kittler, Friedrich, (1990), *Discourse networks 1800/1900*, trans. by Michael Metteer and Chris Cullens, Stanford: Stanford University Press.

Klein, Norman M., (2004), *The Vatican to Vegas: A history of special effects*, New York & London: The New Press.

Kline, Morris, (1954), *Mathematics in Western culture*, London: George Allen and Unwin.

Knight, Frank H., (1921), *Risk, uncertainty, and profit*, Boston & New York: Houghton Mifflin.

Knorr Cetina, Karin, (1999), *Epistemic cultures: How sciences make knowledge*, Cambridge: Harvard University Press.

Koivunen, Niina and Rehn, Alf, (2009), *Creativity and the contemporary economy*, Malmö: Liber.

Kogut, Bruce and Macpherson, J. Muir, (2011), The mobility of economists and the diffusion of policy ideas: The influence of economics on national policies, *Research Policy*, 40: 1307–1320.

Kondo, Dorinne K., (1990), *Crafting selves: Power, gender, and discourses of identity in a Japanese workplace*, Chicago & London: The University of Chicago Press.

Konstantinou, Efrosyni and Fincham, Robin, (2011), Not sharing but trading: Applying Maussian exchange framework to knowledge management, *Human Relations*, 64(6): 823–842.

Korczynski, Marek, (2011), The dialectical sense of humour: Routine joking in a Taylorized factory, *Organization Studies*, *32(*10) 1421–1439.

Koyré, Alexandre, (1959), *From the closed world to the infinite universe*, New York: Harper Torchbooks.

Kracauer, Siegfried, (1995a), Photography, in *The mass ornament: Weimar essays*, trans. by Thomas Y. Levin, Cambridge & London: Harvard University Press, pp. 47–64.

Kracauer, Siegfried, (1995b), The little shopgirls go to the movies, in *The mass ornament: Weimar essays*, trans. by Thomas Y. Levin, Cambridge & London: Harvard University Press, pp. 291–306.

Kracauer, Siegfried, (1995c), George Simmel, in *The mass ornament: Weimar essays*, trans. by Thomas Y. Levin, Cambridge & London: Harvard University Press, pp. 225–257.

Krause, E.A., (1996), *Death of the guilds*, New Haven: Yale University Press.

Krippner, Greta R., (2005), The financialization of the American economy, *Socio-Economic Review*, 3(2): 173–208.

Krippner, Greta R., (2011), *Capitalizing on crisis: The political origins of the rise of finance*, Cambridge & London: Harvard University Press.

Kundera, Milan, (1988), *The art of the novel*, trans. by Linda Asher, New York: Grove Press.

Lakatos, I., (1970), *Proofs and refutations: The logic of mathematical discovery*, Cambridge: Cambridge University Press.

Lakoff, Andrew, (2006), *Pharmaceutical Reason: Knowledge and Value in Global Psychiatry*, Cambridge: Cambridge University Press.

Lakoff, Andrew & Klinenberg, Eric, (2010), Of risk and pork: Urban security and the politics of objectivity, *Theory and Society*, 39: 502–525.

Landes, David, (1983), *Revolution in time: Clocks and the making of the modern world*, Cambridge & London: Harvard University Press.

Latour, Bruno, (1988), *The pasteurization of France*, trans. by Alan Sheridan and John Law, Cambridge & London: Harvard University Press.

Latour, Bruno and Woolgar, S., (1979), *Laboratory life: The construction of scientific facts*, New Jersey: Princeton University Press.

Lawson, M.B. (Buff), (2001), In praise of slack: Time is of the essence, *Academy of Management Executive*, 15(3): 125–135.

Layton, Edwin, ([1971], 1986), *The revolt of the engineers: Social responsibility and the American engineering profession*, Baltimore: John Hopkins University Press.

Lazonick, William, (2010), Innovative business models and varieties of capitalism: Financialization of the U.S. corporation, *Business History Review*, 84: 675–702.

Lazonick, William and O'Sullivan, Mary, (2000), Maximizing shareholder value: a new ideology for corporate governance, *Economy and Society*, 29(1): 13–35.

Lazonick, William and Tulum, Öner, (2011), US biopharmaceutical finance and the sustainability of the biotech business model, *Research Policy*, 40: 1170–1187.

Le Goff, Jacques, (1986/1988), *Your money or your life: Economy and religion in the Middle Ages*, New York: Zone Books.

Lee, Nick and Brown, Steven D., (2002), The disposal of fear. Childhood, trauma and complexity, in Law, John and Mol, Annemarie, eds., (2002), *Complexities: Social studies of knowledge practices*, Durham & London: Duke University Press.

Lee, Richard A. Jr., (2007), Politics and the thing: Excess as the matter of politics, in Winnubst, Shannon, ed., (2007), *Reading Bataille now*, Bloomington & Indianapolis: Indiana University Press, pp. 240–251.

Lefebvre, Henri, (1991), *The production of space*, Oxford: Blackwell.

Lemke, Thomas, (2011), Critique and experience in Foucault, *Theory, Culture and Society*, 28(4): 26–48.

Leonardi, Paul M., (2011), Innovation blindness: Culture, frames, and cross-boundary problems conduction in the development of new technology concepts, *Organization Science*, 22(2): 347–369.

Leroi-Gourhan, André, (1993), *Gesture and speech*, Cambridge and London: The MIT Press.

Lévi-Strauss, Claude, (1985), *The view from afar*, trans. by Joachim Neugroshel and Phoebe Hoss, Oxford. Blackwell.

Lichtenthaler, Ulrich, (2011), Open Innovation: Past research, current debates, and future directions, *Academy of Management Perspectives*, 25(1): 75–93.

Lindenberg, Siegwart and Frey, Bruno S., (1993), Alternatives, frames, and relative prices: A broader view of rational choice theory, *Acta Sociologica*, 36: 191–205.

Lingard, Helen and Francis, Valerie, (2004), The work-life experiences of office and site-based employees in the Australian construction industry, Construction *Management & Economics*, 22(9): 991–1002.

Linstead, Stephen, (2002), Organizational kitsch, *Organization*, 9(4): 657–682.

Lipovetsky, Gilles, ([1994], 2007), A century of fashion, in Barnard, Malcolm, ed., (2007), *Fashion theory: A reader*, London & New York: Routledge, pp. 76–84.

Liu, Alan, (2004), *Laws of the cool: Knowledge work and the culture of information*, Chicago & London: The University of Chicago Press.

Livingston, Eric, (1986), The *ethnomethodological foundations of mathematics*, London, Boston & Henley: Routledge & Kegan Paul.

Lock, Margaret, (2002), *Twice Dead: Organ Transplants and the Reinvention of Death*, Berkeley, Los Angeles & London: University of California Press.

Lord, Richard A. and Siato, Yoshie, (2010), Trends in CEO compensation and equity holdings for S&P 1500 firms: 1994–2007, Journal of Applied Finance, 3(2): 40–56.

Lounsbury, Michael, (2008), Institutional rationality and practice variation: New directions in the institutional analysis of practice, *Accounting, Organization and Society*, 33: 349–361.

Lounsbury, Michael and Crumley, Ellen T., (2007), New practice creation: An institutional approach to innovation, *Organization Studies*, 28: 993–1012.

Love, E. Geoffrey and Nohria, Nitin, (2005), Reducing slack: The performance consequences of downsizing by large industrial firms, 1977–1993, *Strategic Management Journal*, 26(12): 1087–1108.

Luhmann, Niklas, (2000), *The reality of the mass media*, Cambridge: Polity Press.

Lymer, Gustav, (2009), Demonstrating professional vision: The work of critique in architctural education, *Mind, Culture and Activity*, 16: 145–171.

Lynch, Michael, (1991), Pictures of nothing? Visual construals in social theory, *Sociological Theory*, 9(1): 1–21.

Mackenzie, Donald, (1993), Negotiating arithmetic, constructing proof: The sociology of mathematics and information technology, *Social Studies of Science*, 23: 37–65.

Mackenzie, Donald, (1999), Slaying the Kraken: The sociohistory of a mathematical proof, *Social Studies of Science*, 29(1): 7–60.

Mackenzie, Donald, (2004), The big, bad wolf and the rational market: Portfolio investment, the 1987 crash and the performativity of economics, *Economy and Society*, 33(3): 303–334.

Mackenzie, Donald, (2006), *An engine, not a camera: How financial models shape markets*, Cambridge & London, The MIT Press.

Mackenzie, Donald, (2007), The material production of virtuality: Innovation, cultural geography and facticity in derivatives markets, *Economy and Society*, 36(3): 355–376.

Mackenzie, Donald, (2009), *Material markets: How economic agents are constructed*, Oxford & New York: Oxford University Press.

MacKenzie, Donald, (2011), The credit crisis as a problem in the sociology of knowledge, *American Journal of Sociology*, 116(6): 1778–1841.

Mackenzie, Donald and Millo, Yuval, (2003), Constructing a market. Performing a theory: A historical sociology of a financial market derivatives exchange, *American Jounal of Sociology*, 109(1): 107–145.

Madrick, Jeff, (2011), *Age of greed: The triumph of finance and the decline of America, 1970 to the present*, New York: Alfred A. Knopf.

Manovich, Lev, (2001), *The language of new media*, Cambridge & London: The MIT Press.

Manovich, Lev, (2009), The practice of everyday (media) life: From mass consumption to mass cultural production? *Critical Inquiry*, 35 (Winter): 319–331.

Maravelias, Christian, (2003), Post-bureaucracy: Control through professional freedom, *Journal of Organizartion Change Management*, 16(5): 547–566.

March, James G., (1991a), How decisions happen in organizations, *Human-Computer Interactions*, 6: 95–117.

March, James G., (1991b), Exploration and exploitation in organizational learning, *Organization Science*, 2(1): 71–87.

March, James G. and Simon, Herbert A., (1958), *Organizations*, 2nd ed., Blackwell, Oxford.

Maurer, Indre and Ebers, Mark, (2007), Dynamics of social capital and their performance implications: Lessons from biotechnology start-ups, *Administrative Science Quarterly*, 52: 262–292.

Mauss, Marcel, (1954), *The Gift: Forms and Functions of Exchanges in Archaic Societies*, Routledge and Kegan Paul, London.

McAfee, Kathleen, (2003), Neoliberalisms on the molecular scale. Economic and genetic reductionism in the biotechnology battles, *Geoforum*, 34(2): 203–219.

McCloskey, Deirdre N., (1986), *The Rhetorics of Economics*, Wheatsheaf, Brighton.

McCloskey, Deirdre N., (2006), *The bourgeois virtues: Ethics for an age of commerce*, Chicago & London: The University of Chicago Press.

McGirr, Lisa, (2001), *Suburban warriors: The origins of the new American right*, Princeton: Princeton University Press.

McGirr, Lisa, (2002), A history of the conservative movement from the bottom up, *Journal of Policy History*, 14(3): 331–339.

McGivern, Gerry and Ferlie, Ewan, (2007), Playing tick-box games: Interrelating defences in professional appraisal, *Human Relations*, 60(9): 1361–1385.

McWhorter, Ladelle, (2007), The private life of birds: From a restrictive to a general economy of reason, in Winnubst, Shannon, ed., (2007), *Reading Bataille now*, Bloomington & Indianapolis: Indiana University Press, pp. 141–166.

Mellahi, Kamel and Wilkinson, Adrian, (2010), A study of the association between level of slack reduction following downsizing and innovation output, *Journal of Management Studies*, 47(3): 483–508.

Merton, Robert K., (1973), *The sociology of science: Theoretical and empirical investigations*, ed. by Norman W. Storer, Chicago: The University of Chicago Press.

Miles, Raymond E., Snow, Charles C. and Miles, Grant, (2007), The ideology of innovation, *Strategic Organization*, 5(4): 423–435.

Miller, Danny, ed., (1995), *Acknowledging consumption: A review of new studies*, London & New York: Routledge.

Miller, Michael, (1981), *The Bon Marché: The Bourgeois culture and the department store, 1869–1920*, Princeton: Princeton University Press.

Mills, C. Wright, (1959), *The sociological imagination*, Oxford: Oxford University Press.

Milonakis, Dimitris and Fine, Ben, (2009), *From political economy to economics: Method, the social and the historical in the evolution of economic theory*, London & New York: Routledge.

Mirowski, Philip and Van Horn, Robert, (2005), The contract research organization and the commercialization of scientific research, *Social Studies of Science*, 35(4): 503–548.

Mitchell, Robert and Waldby, Catherine, (2001), National biobanks: Clinical labor, risk production and the creation of biovalue, *Science, Technology & Human Values*, 35(2): 330–355.

Mizruchi, Mark, (2004), Berle and Means revisited: The governance and politics of large U.S. corporations, *Theory and Society*, 33: 519–617.

Mizruchi, Mark S. and Brewster, Linda Stearns, (2005), Banking and financial markets, in Smelser, Neil J. and Swedberg, Richard, (2005), *The handbook of economic sociology*, 2nd ed., Princeton & London: Princeton University Press, pp. 284–306.

Mizruchi, Mark and Kimeldorf, Howard, (2005), The historical context of shareholder value capitalism, *Political Power and Social Theory*, 17: 213–221.

Moe, Terry M., (1984), The new economics of organization, *American Journal of Political Science*, 28: 739–777.

Monaghan, Lee F., Hollands, Robert and Pritchard, Gary, (2010), Obesity epidemic entrepreneurs: Types, practices and interests, *Body & Society*, 16(2): 13–71.

Mondzain, Marie-José, (2005), *Image, icon, economy: The Byzantine origin of the contemporary imaginary*, Stanford: Stanford University Press.

Mouritsen, Jacob, Larsen H.T. and Bukh, P.N.D., (2001), Intellectual capital and the 'capable firm': Narrating, visualising and numbering for management knowledge, *Accounting, Organization and Society*, 26: 735–762.

Mumford, Lewis, (1934), *Technics and Civilization*, San Diego, New York & London: Harcourt Brace Jovanovich.

Munck, Ronaldo, (2005), Neoliberalism and politics, and the politics of neoliberalism, in Saad-Filho, Alfredo and Johnston, Deborah, eds., (2005), *Neoliberalism: A critical reader*, London & Ann Arbor: Pluto, pp. 60–69.

Murningham, J. Keith and Conlon, Donald E., (1991), The dynamics of intense work groups: A study of British String quartets, *Administrative Science Quarterly*, 36: 165–186.

Murray, Fiona, (2002), Innovation as co-evuluation of scientific and technological networks: Exploring tissue economics, *Research Policy*, 31: 1389–403.

Myers, Natasha, (2008), Molecular embodiments and the body-work of modeling in protein crystallography, *Social Studies of Science*, 38: 163–199.

Nahapiet, Janine and Ghosal, Sumantra, (1998), Social capital, intellectual capital, and the organizational advantage, *Academy of Management Review*, 23(2): 242–266.

Nayak, Ali, (2008), On the way to theory: A processual appraoch, *Organization Studies*, 29(2): 173–190.

Neal, Mark, (2005), 'I lose, but that is not the point': Situated economic and social rationalities in horserace gambling, *Leisure Studies*, 24(3): 291–310.

Nietzsche, Fridrich, (1976), Twilight of the idols, in *The Portable Nietzsche*, Kaufmann, Walter, ed. (1976), London: Penguin.

Nightingale, Paul, (1998), A cognitive model of innovation, *Research Policy*, 27: 698–709.

Nohria, Nitin and Gulati, Ranjay, (1996), Is slack good for innovation?, *Academy of Management Journal*, 39(5): 1245–1264.

Nord, Walter R., (1992), Alvin W. Gouldner as Intellectual Hero, *Journal of Management Inquiry*. 1992; 1: 350–355.

Noys, Benjamin, (2000), *George Bataille*, London: Pluto Press.

Nussbaum, Martha C., (2010), *Not for profit: Why democracy needs the humanities*, Princeton: Princeton University Press.

Näslund, B., (1964), Organizational slack, *Ekonomisk Tidskrift*, 66(1): 26–31.

Orlikowski, Wanda J., (2010), The sociomateriality of organizational life: Considering technology in management research, *Cambridge Journal of Economics* (34): 125–141.

Oudshoorn, Nelly and Pinch, Trevor, eds., (2003), *How users matter. The co-construction of users and technologies*, Cambridge & London; The MIT Press.

Padgett, John F. and Ansell, Christopher K., (1993), Robust action and the rise of the Medici, 1400–1434, *American Journal of Sociology*, 98: 1259–1319.

Pande, Amrita, (2009), "It may be her eggs but it's my blood." Surrogates and everyday forms of kinship in India, *Qualitative Sociology*, 32: 379–397.

Pearce, Fred, (2003), Introduction: The Collège de sociologie and French social thought, *Economy & Society*, 32(1): 1–6.

Pedraza, Silvia, (2002), A sociology of our times: Alvin Gouldner's message, *The Sociological Quarterly*, 43(1): 73–79.

Peet, Richard, (2007), *Geography of power. The making of global economic policy*, London & New York. Zed Books.

Perlow, Leslie A., (1998), Boundary Control: The Social Ordering of Work and Family Time in High-tech Corporation, *Administrative Science Quarterly*, vol. 43(2): 328–357.

Perlow, Leslie A., (1999), The time famine; Toward a sociology of work time, *Administrative Science Quarterly*, 44: 57–81.

Perrow, Charles, (1986), Economic theories of organization, *Theory and Society*, 15(1/2): 11–45.

Petroski, Henry, (1996), *Inventing by design: How engineers get from thought to thing*, Cambridge: Harvard University Press.

Petryna, Adriana, (2009), *When experiments travel: Clinical trials and the global search for human subjects*, Durham & London: Duke University Press.

Pfeffer, Jeffrey, (1993), Barriers to the advance of organizational science: Paradigm development as a depandent variable, *Academy of Management Review*, 18(4): 599–620.

Pickering, Andrew, (1995), *The mangle of practice: Time, agency, and science*, Chicago & London: The University of Chicago Press.

Pickering, Andrew, (2010), *The cybernetic brain: Sketches of another future*, Chicago & London: The University of Chicago Press.

Pinch, Trevor and Trocco, Frank, (2002), *Analog days: The invention and impact of the Moog synthesizer*, Cambridge & London: Harvard University Press.

Pitts-Taylor, Victoria, (2007), *Surgery junkies: Wellness and pathology in cosmetic culture*, New Brunswick: Rutgers University Press.

Pitts-Taylor, Victoria, (2010), The plastic brain: Neoliberalism and the neuronal self, *Health*, 14(6): 635–652.

Plato, (1995), *Phaedrus*, Indianapolis and Cambridge: Hackett Publishing.

Plehwe, Dieter, Walpen, Bernhard and Neunhöffer, Gisela, (2006), Introduction: Reconsidering neoliberal hegemony, in Plehwe, Dieter, Walpen, Bernhard and Neunhöffer, Gisela, eds., (2006), *Neoliberal hegemony: A global critique*, New York & London: Routledge, pp. 1–24.

Podolny, Joel M., (1993), A status-based model of market compensation, *American Journal of Sociology*, 98: 829–872.

Podolny, Joel M., (1994), Market uncertainty and the social character of economic exchange, *Administrative Science Quarterly*, 39: 458–483.

Podolny, Joel M. and Hill-Popper, Marya, (2004), Hedonic and transcendent conceptions of value, *Industrial and Corporate Change*, 13: 61–89.

Popper, Karl R., (1959), *The Logic of Scientific Discovery*, London: Hutchinson.

Porter, Theodore M., (1995a), *Trust in numbers: The pursuit of objectivity in science and public life*, Princeton: Princeton University Press.

Porter, Theodore M., (1995b), Precision and trust: Early Victorian insurance and the politics of calculation, in Wise, M. Norton, ed., (1995), *The values of precision*, Princeton: Princeton University Press, pp. 173–197.

Porter, Theodore M., (2009), How science became technical, *Isis*, 100: 292–309.

Portes, Alejandro, (1998), Social capital: Its origin and applications in modern sociology, *Annual Review of Sociology*, 23: 1–27.

Powell, Walter W., Koput, Kenneth W. and Smith-Doerr, Laurel, (1996), Interorganizational collaboration and the locus of innovation: Networks of learning in biotechnology, *Administrative Science Quarterly*, 41: 116–145.

Prasad, Anshuman and Prasad, Pushkala, (2003), The postcolonial imagination, in Prasad, Anshuman, ed., (2003), *Postcolonial theory and organizational analysis: A critical engagement*, Houndsmills & New York: Palgrave (pp. 283–297).

Prasad, Pushkala and Prasad, Anshuman, (2002), Casting the native subject: Etnographic practice and the (re)production of difference, in Czarniawska, Barbara and Höpfl, Heather, eds., (2002), *Casting the other: The production and maintenance of inequalities in work organizations*, London & New York: Routledge.

Pratt, M.G., Rockmann, K.W. and Kaufman, J.B., (2006), Constructing professional identity: The role of work and identity learning cycles in the customization of identity among medical residents, *Academy of Management Journal*, 49: 235–262.

Pulley, Thomas I., (2005), From Keynesianism to neoliberalism: Shifting paradigms in economics, in Saad-Filho, Alfredo and Johnston, Deborah, eds., (2005), *Neoliberalism: A critical reader*, London & Ann Arbor: Pluto, pp. 20–29.

Quattrone, Paolo, (2006), The possibility of testimony: A case of case study research, *Organization*, 13(1): 143–157.

Radcliffe-Brown, A.R., (1958), *Methods in social anthropology*, Chicago: The University of Chicago Press.

Raelin, Joseph A., (1985), *The clash of cultures: Managers and professionals*, Boston: Harvard Business School Press.

Ramirez, Paulina and Tylecore, Andrew, (2004), Hybrid corporate governance and its effects innovation. A case study of AstraZeneca, *Technology Analysis & Strategic Management*, 16(1): 97–119.

Rao, Haygreeva, (1994), The social construction of reputation; Certification contests, legitimation, and the survical of organizations in the American automobile industry 1985–1912, *Strategic Management Journal*, 15: 29–44.

Rappaport, Erika D., (1995), 'A new era of shopping': The promotion of women's pleasure in London's West End, 1909–1914, in Carney, Leo and Schwartz, Vanessa R., eds., (1995), *Cinema and the invention of modern life*, Berkeley: University of California Press, pp. 130–155.

Rehn, Alf, (2001), *Electronic potlatch*, Ph.D. Diss., Dept. of Industrial Management, Stockholm: Royal Institute of Technology.

Rehn, Alf and O'Doherty, Damian, (2007), Organization: On the theory and practice of excess, *Culture and Organization*, 13(2): 99–113.

Reichenbach, Hans, (1938), *Experience and prediction*, Chicago: Chicago University Press.

Rheinberger, Hans-Jörg, (1997), *Toward a history of epistemic things: Synthesizing proteins in the test tube*, Stanford: Stanford University Press.

Rheinberger, Hans-Jörg, (2010), *The epistemology of the concrete: Twentieth-century histories of life*, Durham & London: Duke University Press.

Rhodes, Carl, (2001), *Writing organization*, Amsterdam & Philadelpia: John Benjamins.

Richman, Michèle, (2003), Myth, power and the sacred. Anti-utilitarianism in the Collège de sociologie 1937–9, *Economy & Society*, 32(1): 29–47.

Richtnér, Anders and Åhlström, Pär, (2010), Organizational slack and knowledge creation in product development projects: The role of project deliverables, *Creativity and Innovation Management*, 19(4): 428–437.

Richtnér, Anders and Åhlström, Pär, (2006), Influences of organizational slack in new product development projects, *International Journal of Innovation Management*, 10(4): 375–406.

Robertson, Maxine and Swan, Jacky, (2004), Going public: the emergence and effects of soft bureaucracy within a knowledge-intensive firm, *Organization*, 11(1): 123–148.

Robson, Keith, (1992), Accounting numbers as inscriptions: Action at distance and the development of accounting, *Accounting, Organizations, and Society*, 17(7): 685–708.

Romero, Eric and Pescosolido, Anthony, (2008), Human and group effectiveness, *Human Relations*, 6(3): 395–418.

Roper, Juliet, Ganesh, Shiv and Inkson, Kerr, (2010), Neoliberalism and knowledge interests in boundaryless careers discourse, *Work, Employment and Society*, 24(4): 661–679.

Rosen, Michael, (1985), Breakfast at Sprio's: dramaturgy and dominance, *Journal of Management*, 11(2): 31.48.

Rosen, Michael and Astley W.G., (1988), Christmas Time and Control: An Exploration in the Social Structure of Formal Organizations, *Research in the Sociology of Organizations*, vol. 6, pp. 159–182.

Roth, Wolff-Michael, (2009), Radical uncertainty in scientific discovery work, *Science, Technology & Human Values*, 34(3): 313–336.

Roth, Wolff-Michael and Bowen, G. Michael, (1999), Digitalizing lizards: The topology of 'vision' in ecological fieldwork, *Social Studies of Science*, 29(5): 719–764.

Rotman, Brian, (1987), *Signifying nothing: The semiotics of zero*, New York: St. Martin's Press.

Roy, Donald, (1952), Quote restriction and goldbricking in a machine shop, American Journal of Sociology, 57(5): 427–442.

Rushkoff, Douglas, (2009), *Life, Inc.: How the world became a corporation and how to take it back*, London: Vintage.

Saad-Filho, Alfredo and Johnston, Deborah, (2005), Introduction, in Saad-Filho, Alfredo and Johnston, Deborah, eds., (2005), *Neoliberalism: A critical reader*, London & Ann Arbor: Pluto, pp. 3–6.

Sahlins, Marshall, (1972), *Stone age economics*, Chicago & New York: Aldine-Atherton.

Sahlins, Marshall, (2000), The original affluent society, in *Culture in practice: Selected essays*, New York: Zone Books.

Sahlins, Marshall, (2010), Infrastructuralism, *Critical Inquiry*, 36: 371–385.

Sanders, Teela, (2004), Controllable laughter: Managing sex work through humour, *Sociology*, 38(2): 273–291.

Sapir, Edward, ([1931], 2007), Fashion, in Barnard, Malcolm, ed., (2007), *Fashion theory: A reader*, London & New York: Routledge, pp. 37–45.

Sassatelli, Roberta, (2011), Interview with Laura Mulvey: Gender, gaze and technology in film culture, *Theory, Culture & Society*, 28(5): 123–143.

Saxenian, AnnaLee, (1994), *Regional advantage: Culture and competition in Silicon valley and Route 128*, Cambridge & London: Harvard University Press.

Schaffer, Simon, (1995), Accurate measurement is an English Science, in Wise, M. Norton, ed., (1995), *The values of precision*, Princeton: Princeton University Press, pp. 135–172.

Schiller, Friedrich, (1795/2004), *On the aesthetic education of man*, Mineola: Dover Publications.

Schleef, Debra J., (2006), *Managing elites: Professional socialization in law and business schools*, Lanham: Rowman & Littlefield.

Schor, Juliet B., (1993), *The Overworked American*, New York: Basic Books.

Schumpeter, Joseph A., (1942), *Capitalism, Socialism, and Democracy*, Harper & Row, New York.

Schüll, Natasha Dow and Zaloom, Caitlin, (2011), The shortsighted brain: Neuroeconomics and the governance of choice in time, *Social Studies of Science*, 41(4):515–538.

Sconce, Jeffrey, (2000), *Haunted media: Electronic presence from telegraphy to television*, Durham & London: Duke University Press.

Scott, Richard W., (2004), Reflections on a half-century of organizational sociology, *Annual Review of Sociology*, 30: 1–21.

Selin, Cynthia, (2007), Expectations and the emergence of nanotechnology, *Science, Technology & Human Values*, 32(2): 196–220.

Semadeni, Matthew and Anderson, Brian S., (2010), The follower's dilemma: Innovation and imitation in the professional services industry, *Academy of Management Journal*, 53(5): 1175–1193.

Serres, Michel, (1982), *The parasite*, Baltimore: John Hopkins University Press.

Serres, Michel, (1995), *Genesis*, Ann Arbor: University of Michigan Press.

Sewell, William H. Jr., (1992), A theory of structure: Duality, agency, and transformation, *American Journal of Sociology*, 98: 1–29.

Sharp, Lesley A., (2000), The commodification of the body and its parts, *Annual Review of Anthropology*, 29: 287–328.

Sharp, Lesley A., (2011), The invisible woman: The bioaestetics of engineered bodies, *Body & Society*, 17(1): 1–30.

Shenhav, Yehouda, (1999), *Manufacturing rationality. The engineering foundation of the managerial revolution*, Oxford & New York: Oxford University Press.

Shostak, Sara and Conrad, Peter, (2008), Sequencing and its consequences: Path dependence and the relationships between genetics and medicalization, *American Journal of Sociology*, 114: S287–S316. (Special issue on biomedicalization)

Simmel, Georg, (1971), *On individuality and social forms, Selected writings*, Levine, D.A., ed., Chicago: The University Press of Chicago.

Simon, Herbert A., (1957), *Models of Man*, Wiley, New York.

Singer, Ben, (1995), Modernity, hyperstimulus, and the rise of the popular sensationalism, in Carney, Leo and Schwartz, Vanessa R., eds., (1995), *Cinema and the invention of modern life*, Berkeley: University of California Press, pp. 72–99.

Sklair, Leslie, (2002), *The transnational economic class*, London & New York: Routledge.

Slappendel, Carol, (1996), Perspectives on innovation in organizations, *Organization Studies*, 17(1): 107–129.

Sontag, Susan, (1973), *On photography*, New York: Farrar, Straus & Giroux.

Sorkin, Andrew Ross, (2009), *Too big to fail: The insider story of how Wall Street and Washington fought to save the financial system – and themselves*, New York: Viking.

Stafford, Barbara Maria, (2009), Thoughts not our own: Whatever happened to selective attention, *Theory, Culture & Society*, 26(2–3): 275–293.

Standage, Tom, (1998), *The Victorian internet: The remarkable story of the telegraph and the nineteenth century's online pioneers*, London: Weidenfeld & Nicolson.

Star, Susan Leigh, (1999), The ethnography of infrastructure, *American Behavioral Scientist*, 43(3): 377–391.

Starbuck, William H., (2004), Why I stopped trying to understand the real world, *Organization Studies*, 25(7): 1233–1254.

Stark, David, (2009), *The sense of dissonance: Accounts of worth in economic life*, Princeton: Princeton University Press.

Stearns, Linda Brewster, (1986), Capital market effects on external control of corporations, *Theory and Society*, 15(1/2): 47–75.

Stiglitz, Joseph E., (2010a), The financial crisis of 2007–2008 and its macroeconomic consequences, in Griffith-Jones, Stephany, Ocampo, José Antonio and Stiglitz, Joseph E., eds., (2010), *Time for a visible hand: Lessons from the 2008 world financial crisis*, Oxford & New York: Oxford University Press.

Stiglitz, Joseph E., (2010b), *Freefall: America, free markets, and the sinking of the world economy*, New York & London: W.W. Norton.

Styhre, Alexander, (2008), The play of innovation: New drug development and Roger Caillois's theory of play, *Creativity and Innovation Management*, 17(2): 136–146.

Subramaniam, Mohan and Youndt, Mark A., (2005), The influence of intellectual capital on the types of innovative capabilities, *Academy of Management Journal*, 48(3): 450–463.

Suchman, Lucy A., (2007), *Human-machine reconfigurations: Plans and situated actions*, Cambridge: Cambridge University Press.

Suddaby, Roy, Hardy, Cynthia, Huy, Quy Nguyen, (2011), Where are the new theories of organization?, *Academy of Management Review*, 36(2): 236–246.

Sunder Rajan, Kaushik, (2006), *Biocapital: The Constitution of Postgenomic Life*, Durham: Duke University Press.

Swedberg, Richard, (2005), Markets in sociology, in Smelser, Neil J. and Swedberg, Richard, (2005), *The handbook of economic sociology*, 2nd ed., Princeton & London: Princeton University Press, pp. 233–253.

Swedberg, Richard, (2012), Theorizing in sociology and social science: Turning to the context of discovery, *Theory and Society*, 41: 1–40.

Sørensen, Bent Meier and Spoelstra, Sverre, (2012), Play at work: continuation, intervention and usurpation, *Organization*, 19(1) 81–97.

Tarde, Gabriel, ([1890], 1962), *The laws of imitation*, trans. by Elsie Clews Parsons, Gloucester, MA: Peter Smith.

Teman, Elly, (2010), *Birthing a mother: The surrogate body and the pregnant self*, Berkeley, Los Angeles and London: University of California Press.

ten Bos, René, (2000), *Fashion and utopia in management thinking*, Amsterdam & Philadelpia: John Benjamins.

Terrion, Jenepher Lennox and Ashforth, Blake E., (2002), From 'I' to 'we': the role of putdown humour and identity in the development of a temporary group, *Human Relations*, 55(1): 55–88.

Thacker, Eugene, (2006), *The Global Genome: Biotechnology, Politics and Culture*, Cambridge & London: MIT Press.

Thacker, Eugene, (2010), *After life*, Chicago & London: The University of Chicago Press.

Theodosius, Catherine, (2006), Recovering emotions from emotion management, *Sociology*, 40(5): 893–910.

Thomas, Douglas, (2006), *Hacker culture*, Minneapolis & London: The University of Minnesota Press.

Tomaskovic-Devey, Donald and Lin, Ken-Hou, (2011), Income dynamics, economic rents, and the financialization of the U.S. economy, *American Sociological Review*, 76(4) 538–559.

Thompson, Eric P., (1967), Time, work-discipline, and industrial capitalism, *Past & Present*, 38: 56–97.

Thornton, Patricia H., (1999), The sociology of entrepreneurship, *Annual Review of Sociology*, 25: 19–46.

Throsby, Karen, (2009), The war on obesity as a moral project: Weight loss drugs, obesity surgery and negotiating failure, *Science as Culture*, 18(2): 210–216.

Titmuss, Richard M., (1970), *The gift relationship: From human blood to social policy*, London: George Allen & Unwin.

Tiles, Mary, (1991), *Mathematics and the image of reason*, London & New York: Routledge.

Toker, Umut and Gray, Denis O., (2008), Innovation spaces. Workspace planning and innovation in U.S. university research centers, Research Policy, 37: 309–329.

Tolbert Pamela S. and Zucker, Lynne G., (1983), Institutional sources of change in the formal structure of organizations; The diffusion of civil service reform, 1880–1935, *Administrative Science Quarterly* 30: 2239.

Torgersen, Helge, (2009), Fuzzy genes: Epistemic tensions in genomics, *Science as Culture*, 18(1): 65–87.

Townley, Barbara, (2008), *Reason's neglect: Rationality and organzing*, Oxford & New York: Oxford University Press.

Townley, Barbara, Beech, Nic and McKinlay, Alan, (2009), Managing in the creative industries: Managing the motley crew, *Human Relations*, 62(7): 939–962.

Traweek, Sharon, (1988), *Beamtimes and lifetimes: the world of high eneregy physicists*, Cambridge & London: Harvard University Press.

Tsang, Eric W.K. and Zahra, Shaker A., (2008), Organizational unlearning, *Human Relations*, 61(10): 1435–1462.

Turkle, Sherry, (2011), *Alone together*, New York: Basic Books.

Turner, Victor, (1982), *From ritual to theatre: The human seriousness of play*, New York: PAJ Publications.

Tyler, Melissa and Abbott, Pamela, (1998), Chocs away: Weight watching in the contemporary airline industry, *Sociology*, 32(3): 433–450.

Tyler, Melissa and Taylor, Steve, (1998), The exchange of aesthetics: Women's work and 'the gift', *Gender, Work and Organization*, 5(3): 165–171.

Tyler, Tom R., (2011), *Why people cooperate: The role of social motivations*, Princeton: Princeton University Press.

Utterback, James M., (1994), *Mastering the dynamics of innovation*, Boston: Harvard Business School Press.

Uzzi, Brian, (1999), Embeddedness in the making of financial capital: How social relations and networks benefit Firms seeking financing, *American Sociological Review*, 64(4): 481–505.

Van de Ven, Andrew, Angle, Harold L. and Poole, Marshall Scott, eds., (2000), *Research on the management of innovation*, Oxford & New York: Oxford University Press.

Vasari, Giorgio, ([1550], 1991), *The lives of the artists*, trans. by Julia Conway Bondanella and Peter Bondanella, Oxford and New York: Oxford University Press.

Veblen, Thorstein, (1899/1994), *The Theory of the Leisure Class*, Penguin, London.

Velthuis, Olav, (2003), Symbolic meanings of prices: Constructing the value of contemporary art in Amsterdam and New York, *Theory and Society*, 32: 181–215.

Virilio, Paul, (2002), *Desert screen: War at the speed of light*, London & New York: Continuum.

Vrecko, Scott, (2008), Capital ventures into biology: Biosocial dynamics in the industry and science of gambling, *Economy and Society*, 37(2): 50–67.

Vrecko, Scott, (2010), Global and everyday matters of consumption: On the productive assemblage of pharmaceuticals and obesity, *Theory and Society*, 39: 555–573.

Vygotsky, Lev, (1978), *Mind in society: The development of higher psychological processes*, edited by Michael Cole, Vera John-Steiner, Sylvia Scribner and Ellen Souberman, Cambridge & London: Harvard University Press.

Wacquant, Loïc J. D., (2009), *Punishing the poor: The neoliberal government of social insecurity*, Durham & London: Duke University Press.

Waguespack, David M. and Sorenson, Olav, (2011), The rating game: Asymmetry in classification, *Organization Science*, 22(3): 541–553.

Wajcman, Judy and Rose, Emily, (2011), Constant connectivity: Rethinking interruptions at work, *Organization Studies*, 32(7): 941–961.

Warhurst, Chris and Nickson, Dennis, (2007), Employee experience of aesthetic labour in retail and hospitality, *Work, Employment and Society*, 21: 103–120.

Warhurst, Chris and Nickson, Dennis, (2009), Who's got the look? Emotional, aesthetic and sexualized labour in interactive services, *Gender, Work and Organization*, 16(3): 385–404.

Wark, McKenzie, (2004), *A hacker manifesto*, Cambridge & London: Harvard University Press.

Warren, Sam and Fineman, Stephen, (2007), 'Don't get me wrong, it's fun here, but …': Ambivalence and paradox in a 'fun' work environment, in Westwood, Robert and Rhodes, Carl, eds., (2007), *Humour, work and organization*, London & New York: Routledge, pp. 92–112.

Warwich, Andrew, (1995), The laboratory of theory or what's exact about exact science, in Wise, M. Norton, ed., (1995), *The values of precision*, Princeton: Princeton University Press, pp. 311–342.

Watts, Jacqueline H., (2009), 'Allowed into a man's world' meanings of work-life balance: perspectives on women civil engineers as 'minority' workers in construction, *Gender, Work and Organization*, 16(1): 37–57.

Weber, Max, (1962), *Basic concepts in sociology*, London: Citadel Press.

Weber, Max, (1992), *The Protestant Ethic and the Spirit of Capitalism*, Londo: Routledge, London.

Weeks, John, (2004), *Unpopular culture: The rituals of complaints in a British bank*, Chicago: The University of Chicago Press.

Weick, Karl E., (1996), Drop your tools: An allegory for organizational studies, *Administrative Science Quarterly*, 40: 301–313.

Weick, Karl E. and Roberts, Karlene H., (1993), Collective mind in organizations: Heedful interrelating on flight decks, *Administrative Science Quarterly*, 38: 357–381.

Weick, Karl E. and Westley, Frances, (1999), Organizational Learning: Affirming an Oxymoron, in Clegg, Stewart R., Hardy, Cynthia and Nord, Walter R., eds., (1999), *Managing Organizations*, London: Sage.

Wendt, Ron, (1999), Corporate Lampoonery as a hermeneutic sign of our times: Reading 'Dilbert' from critical and Poststructuralist perspectives, *Electronic Journal of Radical Organization Theory*, vol. 5, no. 1, pp. 1–14.

Westwood, Robert and Rhodes, Carl, eds., (2007), *Humour, work and organization*, London & New York: Routledge.

Westwood, Robert, (2007), The staging of humour; Organizing and managing comedy, in Westwood, Robert and Rhodes, Carl, eds., (2007), *Humour, work and organization*, London & New York: Routledge, pp. 271–298.

White, Harrison C., (1981), Where do markets come from?, *American Journal of Sociology*, 87(3): 517–547.

Wilkinson, Stephen, (2006), *Bodies for sale: Ethics and exploitation in the human body trade*, London & New York: Routledge.

Williams, Christine L. and Connell, Catherine, (2010), 'Looking good and sounding right': Aesthetic labor and social inequality in the retail industry, *Work and Occupations*, 37(3): 349–377.

Williams, Rosalind, (1990), *Notes on the underground; An essay on technology, society and imagination*, Cambridge: The MIT press.

Willse, Craig, (2010), Neo-liberal biopolitics and the invention of chronic homelessness, *Economy and Society*, 39(2): 144–184.

Winnicott, Donald Woods, (1971), *Playing and reality*, London: Tavistock.

Wittgenstein, Ludwig, (1953), *Remarks in the foundations of mathematics*: Ed. by Georg Henrik von Wright, Rush Rhees & G. Elizabeth M. Anscombe, Trans. by G. Elizabeth M. Anscombe, Oxford: Blackwell.

Wittgenstein, Ludwig, (1967), *Zettel*, Berkeley and Los Angeles: University of California Press.

Wolfe, Richard A., (1994), Organization innovation: Review, critique and suggested research directions, *Journal of Management Studies*, 31(3): 405–431.

Womack, James, Jones, Daniel T. and Roos, Daniel, (1990), *The machine that changed the world*, New York: Macmillan.

Wright, David, (2005), Mediating production and consumption: Cultural capital and cultural workers, *British Journal of Sociology*, 36: 105–121.

Yakubovich, Valery, Granovetter, Mark and McGuire, Patrick, (2005), Electric charges: The social construction of rate systems, *Theory and Society*, 34: 579–612.

Yakura, Elaine K., (2002a), Billables: The valorization of time in consulting, *American Behavioral Scientist*, 44: 1076–1095.

Yakura, Elaine K., (2002b), Charting time: Timelines as temporal boundary objects, *Academy of Management Journal*, 45(5): 956–970.

Zahara, Shaker A. and Newey, Lance R., (2009), Maximizing the impact of organization science: theory-building at the intersection of disciplines and /or fields, *Journal of Management Studies*, 46(6): 1059–1075.

Zajac, Edward J. and Westphal, James D., (2004), The Social Construction of Stock Market Value: Institutionalization and Learning Perspectives on Stock Market Reactions, *American Sociological Review*, 69: 433–57.

Zald, Meyer N. and Lounsbury, Michael, (2010), The wizards of Oz: Towards an institutional approach to elites, expertise, and command posts, *Organization Studies*, 31(7): 963–996.

Zaloom, Caitlin, (2006), *Out of the pits: Trading and technology from Chicago to London*, Durham & London: Duke University Press.

Zelizer, Vivianne, (2005), *The purchase of intimacy*, Princeton: Princeton University Press.

Zerubavel, Eviatar, (1982), The standardization of time: A sociohistorical perspective, *American Journal of Sociology*, 88: 1–29.

Žižek, Slavoj, (2009), *First as tragedy, then as farce*, New York & London: Verso.

Zorn, Dirk M., (2004), Here a chief, there a chief: The rise of the CFO in the American firm, *American Sociological Review*, 69: 345–364.

Zorn, Dirk, Dobbin, Frank, Dierkes, Julian and Kwok, Man-shan, (2005), Managing investors: How financial markets reshaped the American firm, in Knorr Cetina, Karin and Preda, Alex, eds., (2005), *The sociology of financial markets*, Oxford & New York: Oxford University Press. pp. 269–289.

Zott, Christoph and Huy, Quy Nguyen, (2007), How entrepreneurs use symbolic management to acquire resources, *Administrative Science Quarterly*, 52: 70–105.

Zuckerman, Ezra, (1999), The categorical imperative: Securities analysts and the illegitimacy discount, *American Journal of Sociology*, 104: 1398–1438.

Zuckerman, Ezra, (2004), Structural incoherence and stock market activity, *American Sociological Review*, 69: 405–432.

Index